Robotic Flexible Manipulation and Intelligent Sensing Technologies

Juntian Qu

Robotic Flexible Manipulation and Intelligent Sensing Technologies

Springer

Juntian Qu
Shenzhen International Graduate School
Tsinghua University
Shenzhen, China

ISBN 978-981-95-6231-2 ISBN 978-981-95-6232-9 (eBook)
https://doi.org/10.1007/978-981-95-6232-9

This Springer imprint is published by the registered company Springer Nature Singapore Pte Ltd.
The registered company address is: 152 Beach Road, #21-01/04 Gateway East, Singapore 189721, Singapore

Contents

Introduction

1 Origin of Flexible Manipulators

1.1 Biomimicry and Early Exploration

The concept of underwater soft robots originated from the imitation of marine biological motion mechanisms. In nature, the octopus achieves flexible bending and grasping of its tentacles through the coordinated contraction of muscle fibers. Its "muscular hydrostat" structure (shape deformation under constant volume) provides natural inspiration for flexible mechanical design. In the 1960s, the Massachusetts Institute of Technology (MIT) first attempted to use steel cables to control the opening and closing of mechanical fingers, validating the feasibility of flexible actuation. However, limited by material technology, it could only achieve single-degree-of-freedom motion.

Early explorations in actuation technology primarily focused on pneumatic and hydraulic domains. In the 1990s, Germany's Festo company developed the first biomimetic elephant trunk-like flexible robotic arm based on pneumatic actuation principles. It achieved bending and extension through multi-chamber air pressure regulation, but its structure was bulky and control precision was limited. Around the same time, the OctArm series of robotic arms developed by the University of Pennsylvania adopted a three-chamber symmetric pneumatic structure, achieving simple grasping operations in underwater environments for the first time, forming the prototype of underwater flexible robotic arms.

The core challenge during this stage lay in the compatibility between materials and actuation. The rubber materials used initially lacked sufficient elasticity, while the introduction of metal skeletons compromised compliance. It was not until 1997 that the Mars Pathfinder lander used inflatable soft pads made of electroactive polymers (EAP), demonstrating the application potential of flexible materials in extreme environments and offering new ideas for material selection in underwater equipment.

J. Qu, *Robotic Flexible Manipulation and Intelligent Sensing Technologies*,
https://doi.org/10.1007/978-981-95-6232-9_1

1.2 *From Single Flexibility to Multifunctional Composites*

1.2.1 Widespread Application of Silicone-Based Materials

Silicone, due to its high elasticity, corrosion resistance, and biocompatibility, has become a core material for underwater soft hands. Professor Yan Junjie's team at Harbin Institute of Technology used Ecoflex-0050 silicone to fabricate pneumatic soft actuators. Through uniaxial tensile experiments, they determined the parameters of the Yeoh constitutive model, providing an accurate material model for the mechanical analysis of underwater grasping. Such materials can achieve a maximum end displacement of 108.96 mm under air pressure of 0–50 kPa, with stress far below the destruction threshold (12 MPa), meeting the structural stability requirements for underwater operations.

1.2.2 Integrated Innovation of Smart Materials

The development of electroactive polymers (EAP) has driven innovation in actuation methods. The dielectric elastomer actuator (DEA) developed by a research team at Harvard University can generate strains exceeding 100% under an electric field, with a response speed in the millisecond range, providing a lightweight actuation solution for miniature underwater soft hands.

1.3 *Diversified Development of Actuation Methods*

1.3.1 Engineering Breakthroughs in Pneumatic Actuation

Pneumatic actuation utilizes the compressibility of gas to achieve movement of the robotic arm or soft hand by changing air pressure. It offers advantages such as simple structure, fast response, low cost, and no pollution, making it widely used in the field of underwater soft robots.

Early pneumatic soft robotic arms often used single-chamber or simple multi-chamber structures, with limited control precision and output force. To enhance performance, researchers conducted extensive explorations in structural design, material application, and control methods. In structural optimization, biomimetic design concepts were introduced, simulating biological structures like octopus tentacles and elephant trunks. By cleverly arranging multiple air chambers, more flexible and complex motion modes were achieved. For instance, some new pneumatic soft robotic arms distribute multiple air chambers at specific angles and positions, enabling omnidirectional bending and torsion, allowing them to navigate flexibly through complex underwater environments and perform delicate tasks. This is analogous to the octopus-inspired soft gripper, which, through the coordination of multiple arms, achieves multimodal grasping as well as complex motion patterns like omnidirectional crawling and swimming underwater.

In terms of engineering applications, pneumatically driven underwater soft hands and flexible robotic arms have achieved numerous results. In underwater exploration, some underwater robots equipped with pneumatic flexible robotic arms can operate flexibly in complex seabed terrains, performing non-destructive collection of marine organisms and geological samples. In underwater rescue scenarios, pneumatic soft hands, leveraging their soft characteristics, can flexibly grasp objects in confined spaces, avoiding secondary damage to rescue targets and providing strong support for underwater rescue operations. The detachable soft gripper developed by Professor Xie Guangming's team at Peking University demonstrates the advantages of pneumatic actuation in complex underwater environments, capable of grasping various types of irregular objects.

1.3.2 High-Precision Control of Hydraulic Actuation

Hydraulic actuation transmits pressure through liquid, capable of generating large driving forces. It is suitable for underwater operations requiring high grasping force and load capacity, such as deep-sea salvage and underwater engineering maintenance. However, the complexity of the underwater environment, including pressure changes and water current disturbances, poses severe challenges to the precision and stability of hydraulic actuation systems.

To achieve high-precision control, researchers have continuously innovated in system design and control algorithms. At the system design level, advanced sensor technology is introduced to monitor parameters such as pressure and flow in the hydraulic system, as well as the position and posture of the robotic arm in real-time, constructing closed-loop control systems. High-precision pressure sensors can accurately detect pressure changes in the hydraulic circuit and feed signals back to the controller. The controller then adjusts the output flow and pressure of the hydraulic pump based on preset values, thereby achieving precise control of the robotic arm's movement. The application of hydraulic drive technology in underwater environments and the improvement of its power transmission efficiency and control precision rely on in-depth research into the characteristics of hydraulic components and underwater working conditions.

In practical applications, hydraulically driven underwater soft hands and flexible robotic arms perform excellently in deep-sea operations. In deep-sea salvage missions, they can withstand immense underwater pressure to accurately grasp and transport sunken objects. In underwater pipeline maintenance operations, hydraulically driven soft hands can perform fine operations on pipelines, such as tightening screws and welding pipes, effectively ensuring the smooth progress of underwater engineering.

1.3.3 Synergistic Optimization of Cable-Driven and Hybrid Actuation

Cable-driven actuation transmits force by pulling cables, offering advantages like simple structure and ease of control. However, when used alone, it suffers from issues such as insufficient stiffness and susceptibility to interference. Hybrid actuation combines the advantages of multiple actuation methods to compensate for the deficiencies of single actuation, becoming an effective way to enhance the performance of underwater soft hands and flexible robotic arms.

In terms of cable-driven technology improvements, researchers have enhanced the transmission efficiency and reliability of cable-driven systems by optimizing cable materials and structures. Using high-strength, low-friction cable materials reduces energy loss and cable wear; improving cable winding methods and pulley structures reduces jamming and deviation during transmission, thereby improving motion precision. The application of cable-driven technology in underwater environments requires consideration of cable waterproofing, corrosion resistance, and stress changes under water pressure, all of which are important aspects for improving its performance.

Research on hybrid actuation primarily focuses on the cooperative control of different actuation methods. For example, combining cable-driven with pneumatic or hydraulic actuation utilizes cable drive for precise position control, while pneumatic or hydraulic actuation provides auxiliary power and compliance adjustment. The cable-pneumatic hybrid-driven tapered soft manipulator (TSM) proposed by institutions like Harbin Institute of Technology innovatively uses Bowden tubes as driving tendons, constructing a composite structure combining Bowden tubes and cable tendons. Through the hybrid drive antagonistic principle, it achieves tight integration with pneumatic tapered bellows (PATB). This design endows the manipulator with excellent resistance to stretching, bending, and torsion, greatly enhancing the balance between rigidity and flexibility and achieving high precision with an average error of only 0.17 mm in spatial point positioning tasks. Furthermore, the cable-driven flexible arm based on deep reinforcement learning proposed by Qu Juntian's team at Tsinghua University, by integrating Long Short-Term Memory (LSTM) neural networks with fully connected neural networks, enables the deep reinforcement learning control strategy to consider both historical and current policy relationships, improving the control precision of the flexible arm's end position. This flexible arm can not only perform delicate grasping tasks on land but also conduct operations like sampling in narrow, confined spaces underwater. This research provides new ideas and methods for the intelligent application of cable-driven technology in complex underwater environments.

Cable-driven and hybrid actuation show promising application prospects in underwater operations. In detection and sampling tasks in complex underwater terrains, hybrid-driven flexible robotic arms can flexibly adjust actuation methods according to different operational needs, ensuring both motion precision and sufficient driving force and compliance, thereby improving operational efficiency and success rate.

The actuation methods for underwater soft hands and flexible robotic arms are continuously developing toward diversification and high performance. Pneumatic actuation is making constant breakthroughs in engineering applications, enhancing motion flexibility and control precision through structural and material innovations. Hydraulic actuation, aided by advanced sensors and intelligent algorithms, achieves high-precision control and plays a key role in heavy-load underwater operations. Cable-driven and hybrid actuation, through synergistic optimization, integrate the advantages of multiple actuation methods, providing more reliable solutions for complex underwater tasks.

1.4 Development and Breakthroughs in Basic Motion Control

1.4.1 Early Attempts and Breakthroughs in Pose Control

Initially, pose control of underwater soft hands and flexible robotic arms faced numerous challenges. Traditional pose control methods for rigid robotic arms were difficult to apply directly due to the highly nonlinear, large-deformation characteristics of soft and flexible structures. Early attempts were mostly based on simple open-loop control, using preset drive signals to control the robotic arm's movement. However, due to the lack of real-time feedback, pose accuracy was very low.

As research deepened, scholars began to introduce closed-loop control strategies. Inspired by the octopus's rich behavioral repertoire, Professor Xie Guangming's team at Peking University developed a soft gripper capable of multimodal grasping [1]. The team demonstrated their developed six-arm soft gripper. The device, through the coordinated work of its arms, achieves two functional states: grasping and movement. When mounted on a robotic arm, the gripper can perform multimodal grasping tasks; after detaching from the robotic arm, it can still independently perform movement and grasping operations underwater. In this process, by installing pressure sensors, vision sensors, etc., at the end of the gripper, real-time pose information is acquired and compared with preset targets, and then drive signals are adjusted. This provides a feasible idea for closed-loop pose control of underwater soft hands.

Simultaneously, the multi-channel planar bending soft arm developed by researchers at MIT's Computer Science and Artificial Intelligence Laboratory utilized a vision system to achieve curvature and position closed-loop control, enabling the soft arm to navigate through complex environments and ushering in an era of feedback-based precise pose control [2, 3]. On this basis, advanced algorithms such as PID control were applied to the pose control of underwater soft hands and flexible robotic arms. The PID algorithm adjusts driving parameters in real-time based on the deviation between the current pose and the target pose, as well as the rate of change of the deviation, significantly improving pose control accuracy. In some laboratory environments, the pose positioning accuracy of underwater flexible robotic arms based on PID control can reach the millimeter level.

1.4.2 Piecewise Constant Curvature Stage

Early arm shape control for underwater flexible robotic arms often adopted the piecewise constant curvature assumption. This method divides the flexible robotic arm into several segments, each assumed to be in a constant curvature bending state. The overall arm shape is controlled by controlling the curvature of each segment. This method simplified the control model and reduced computational complexity to some extent [2, 3]. For instance, research from institutions like Shandong University divided a cable-driven flexible robotic arm into multiple segments. By controlling the tension of the cables in each segment, the curvature of that segment was changed, thereby adjusting the arm shape. However, the piecewise constant curvature assumption deviates from the continuous, smooth deformation characteristics of actual flexible robotic arms, leading to suboptimal arm shape control performance in complex motion and high-precision scenarios.

1.4.3 The Rise of Continuum Theory

To address the limitations of piecewise constant curvature, continuum theory gradually became a research hotspot in arm shape control. Continuum theory treats the flexible robotic arm as a continuously deformable body. Based on material mechanics and elasticity theory, it establishes precise continuous models to describe arm shape changes [2, 3]. Professor Wen Li's research team at Beihang University, inspired by the octopus's predatory behavior, constructed a new kinematic modeling method of "bending wave propagation" for an octopus-arm-like system, which falls within the scope of continuum theory [4]. By observing the movement of the slender arms of a biological octopus during predation, the research team achieved fast solving of the kinematics of the flexible robotic arm, with a model solving time of only 3 ms, laying the foundation for real-time arm shape control based on continuum theory. Through this model, the arm shape of the flexible robotic arm under different drives can be predicted and controlled more accurately, enabling it to better adapt to complex underwater environments, such as flexibly grasping objects in confined spaces. In some underwater soft robotic arm designs based on continuum theory, combined with advanced manufacturing technologies like 3D printing, the material distribution and structural parameters of the robotic arm can be precisely controlled, further optimizing arm shape control performance. For example, by using materials with different elastic moduli in different parts, the robotic arm can be made easier to bend or maintain stiffness in specific areas, achieving more complex and precise arm shape changes [2, 3].

1.5 *Diversification and Intelligence of Control strategies*

1.5.1 Diversified Control Strategies

In addition to classic model-based control strategies (such as the aforementioned PID control), various new control strategies have emerged in recent years. Adaptive control strategies have been applied in the control of underwater soft hands and flexible robotic arms. Due to the dynamic changes in the underwater environment (e.g., water currents and pressure changes), adaptive control can automatically adjust control parameters based on real-time environmental information and system status to maintain good control performance. In underwater environments with current disturbances, adaptive control enables the flexible robotic arm to adjust its motion trajectory in real-time to stably grasp objects [2, 3]. Sliding mode control strategies have also attracted attention. Sliding mode control has strong robustness against system parameter variations and external disturbances. By designing switching functions, the system switches motion on different sliding surfaces to achieve precise control. In complex underwater disturbance environments, sliding mode control can ensure stable grasping of irregular objects by underwater soft hands, reducing the impact of disturbances on grasping pose and arm shape [2, 3].

1.5.2 Intelligent Control Trends

With the development of artificial intelligence technology, intelligent control has become an important trend in the control of underwater soft hands and flexible robotic arms. Machine learning algorithms, such as neural networks, are used to establish motion models of robotic arms. Researchers from institutions like Harbin Institute of Technology, inspired by elephant trunks, proposed a design scheme for a cable-pneumatic hybrid-driven tapered soft manipulator (TSM), using deep learning neural network modeling technology to accurately simulate the inverse kinematics performance of the TSM [5]. This research, by constructing a composite structure combining Bowden tubes and cable tendons (BTCT), achieved tight integration of BTCT with pneumatic tapered bellows (PATB), endowing the TSM with excellent resistance to axial stretching, tangential bending, and torsion. By training the neural network with a large amount of experimental data, it can quickly and accurately calculate the required driving parameters based on the target pose and arm shape.

Significant progress has been made in the basic motion control of underwater soft hands and flexible robotic arms regarding pose control, arm shape control, and control strategies. From early simple open-loop control to precise closed-loop pose control, from piecewise constant curvature arm shape assumptions to precise modeling based on continuum theory, and from classical control strategies to diversified and intelligent control development, each breakthrough has promoted advancements in underwater operation technology. However, challenges remain, such as long-term stable control in complex marine environments and further improving control accuracy and efficiency. Future research could focus on multidisciplinary integration, combining new materials, advanced manufacturing technologies, and intelligent algorithms to continuously optimize the basic motion control technology of underwater soft hands and flexible robotic arms and expand their applications in fields such as underwater resource development, marine scientific research, and underwater rescue.

2 Development of Flexible Manipulators

The Development History of Underwater Soft Hands and Flexible Robotic Arms: Origins and Landmark Breakthroughs.

The complexity and uniqueness of the underwater environment impose extremely high demands on operational equipment. The limitations of traditional rigid robotic arms in terms of adaptability, safety, and operational flexibility are becoming increasingly apparent. Underwater soft hands and flexible robotic arms, with their unique compliance, environmental adaptability, and safety in human-robot interaction, have become research hotspots in fields such as marine engineering, underwater rescue, and deep-sea exploration. By mimicking the motion mechanisms of biological flexible limbs (such as octopus tentacles and elephant trunks) and utilizing flexible materials and novel actuation technologies, these systems achieve efficient operation in unstructured environments. This article systematically reviews the origins, landmark breakthroughs, and key technological advancements of underwater soft hands and flexible robotic arms, revealing their developmental trajectory from conceptualization to practical application, and provides an outlook on future research directions.

2.1 Biomimicry and Early Exploration

The concept of underwater soft robots originated from the imitation of marine biological motion mechanisms. In nature, the octopus achieves flexible bending and grasping of its tentacles through the coordinated contraction of muscle fibers. Its "muscular hydrostat" structure (shape deformation under constant volume) provides natural inspiration for flexible mechanical design [6]. In the 1960s, the Massachusetts Institute of Technology (MIT) first attempted to use steel cables to control the opening and closing of mechanical fingers, validating the feasibility of flexible actuation. However, limited by material technology, it could only achieve single-degree-of-freedom motion [7].

Early explorations in actuation technology primarily focused on pneumatic and hydraulic domains. In the 1990s, Germany's Festo company developed the first biomimetic elephant trunk-like flexible robotic arm based on pneumatic actuation principles. It achieved bending and extension through multi-chamber air pressure regulation, but its structure was bulky and control precision was limited [6]. Around the same time, the OctArm series of robotic arms developed by the University of

Pennsylvania adopted a three-chamber symmetric pneumatic structure, achieving simple grasping operations in underwater environments for the first time, forming the prototype of underwater flexible robotic arms [6].

The core challenge during this stage lay in the compatibility between materials and actuation. The rubber materials used initially lacked sufficient elasticity, while the introduction of metal skeletons compromised compliance. It was not until 1997 that the Mars Pathfinder lander used inflatable soft pads made of electroactive polymers (EAP), demonstrating the application potential of flexible materials in extreme environments and offering new ideas for material selection in underwater equipment [8].

2.2 Breakthroughs in Material Technology: From Single Flexibility to Multifunctional Composites

2.2.1 Widespread Application of Silicone-Based Materials

Silicone, due to its high elasticity, corrosion resistance, and biocompatibility, has become a core material for underwater soft hands. Professor Yan Junjie's team at Harbin Institute of Technology used Ecoflex-0050 silicone to fabricate pneumatic soft actuators. Through uniaxial tensile experiments, they determined the parameters of the Yeoh constitutive model, providing an accurate material model for the mechanical analysis of underwater grasping [9]. Such materials can achieve a maximum end displacement of 108.96 mm under air pressure of 0–50 kPa, with stress far below the destruction threshold (12 MPa), meeting the structural stability requirements for underwater operations [9].

2.2.2 Integrated Innovation of Smart Materials

The development of electroactive polymers (EAP) has driven innovation in actuation methods. The dielectric elastomer actuator (DEA) developed by a research team at Harvard University can generate strains exceeding 100% under an electric field, with a response speed in the millisecond range, providing a lightweight actuation solution for miniature underwater soft hands [8].

2.3 Diversified Development of Actuation Methods

2.3.1 Engineering Breakthroughs in Pneumatic Actuation

Pneumatic actuation utilizes the compressibility of gas to achieve movement of the robotic arm or soft hand by changing air pressure. It offers advantages, such as simple structure, fast response, low cost, and no pollution, making it widely used in the field of underwater soft robots.

Early pneumatic soft robotic arms often used single-chamber or simple multi-chamber structures, with limited control precision and output force. To enhance performance, researchers conducted extensive explorations in structural design, material application, and control methods. In structural optimization, biomimetic design concepts were introduced, simulating biological structures like octopus tentacles and elephant trunks. By cleverly arranging multiple air chambers, more flexible and complex motion modes were achieved. For instance, some new pneumatic soft robotic arms distribute multiple air chambers at specific angles and positions, enabling omnidirectional bending and torsion, allowing them to navigate flexibly through complex underwater environments and perform delicate tasks [10]. This is analogous to the octopus-inspired soft gripper, which, through the coordination of multiple arms, achieves multimodal grasping as well as complex motion patterns like omnidirectional crawling and swimming underwater [11].

In terms of engineering applications, pneumatically driven underwater soft hands and flexible robotic arms have achieved numerous results. In underwater exploration, some underwater robots equipped with pneumatic flexible robotic arms can operate flexibly in complex seabed terrains, performing non-destructive collection of marine organisms and geological samples. In underwater rescue scenarios, pneumatic soft hands, leveraging their soft characteristics, can flexibly grasp objects in confined spaces, avoiding secondary damage to rescue targets and providing strong support for underwater rescue operations [12]. The detachable soft gripper developed by Professor Xie Guangming's team at Peking University demonstrates the advantages of pneumatic actuation in complex underwater environments, capable of grasping various types of irregular objects [11].

2.3.2 High-Precision Control of Hydraulic Actuation

Hydraulic actuation transmits pressure through liquid, capable of generating large driving forces. It is suitable for underwater operations requiring high grasping force and load capacity, such as deep-sea salvage and underwater engineering maintenance. However, the complexity of the underwater environment, including pressure changes and water current disturbances, poses severe challenges to the precision and stability of hydraulic actuation systems.

To achieve high-precision control, researchers have continuously innovated in system design and control algorithms. At the system design level, advanced sensor technology is introduced to monitor parameters such as pressure and flow in the hydraulic system, as well as the position and posture of the robotic arm in real-time, constructing closed-loop control systems. High-precision pressure sensors can accurately detect pressure changes in the hydraulic circuit and feed signals back to the controller. The controller then adjusts the output flow and pressure of the hydraulic pump based on preset values, thereby achieving precise control of the robotic arm's movement [13]. The application of hydraulic drive technology in underwater environments and the improvement of its power transmission efficiency and control precision rely on in-depth research into the characteristics of hydraulic components and underwater working conditions [14].

In practical applications, hydraulically driven underwater soft hands and flexible robotic arms perform excellently in deep-sea operations. In deep-sea salvage missions, they can withstand immense underwater pressure to accurately grasp and transport sunken objects. In underwater pipeline maintenance operations, hydraulically driven soft hands can perform fine operations on pipelines, such as tightening screws and welding pipes, effectively ensuring the smooth progress of underwater engineering [14].

2.3.3 Synergistic Optimization of Cable-Driven and Hybrid Actuation

Cable-driven actuation transmits force by pulling cables, offering advantages like simple structure and ease of control. However, when used alone, it suffers from issues such as insufficient stiffness and susceptibility to interference. Hybrid actuation combines the advantages of multiple actuation methods to compensate for the deficiencies of single actuation, becoming an effective way to enhance the performance of underwater soft hands and flexible robotic arms.

In terms of cable-driven technology improvements, researchers have enhanced the transmission efficiency and reliability of cable-driven systems by optimizing cable materials and structures. Using high-strength, low-friction cable materials reduces energy loss and cable wear; improving cable winding methods and pulley structures reduces jamming and deviation during transmission, thereby improving motion precision [15]. The application of cable-driven technology in underwater environments requires consideration of cable waterproofing, corrosion resistance, and stress changes under water pressure, all of which are important aspects for improving its performance [15].

Research on hybrid actuation primarily focuses on the cooperative control of different actuation methods. For example, combining cable-driven with pneumatic or hydraulic actuation utilizes cable drive for precise position control, while pneumatic or hydraulic actuation provides auxiliary power and compliance adjustment. The cable-pneumatic hybrid-driven tapered soft manipulator (TSM) proposed by institutions like Harbin Institute of Technology innovatively uses Bowden tubes as driving tendons, constructing a composite structure combining Bowden tubes and cable tendons. Through the hybrid drive antagonistic principle, it achieves tight integration with pneumatic tapered bellows (PATB). This design endows the manipulator with excellent resistance to stretching, bending, and torsion, greatly enhancing the balance between rigidity and flexibility and achieving high precision with an average error of only 0.17 mm in spatial point positioning tasks [11]. Furthermore, the cable-driven flexible arm based on deep reinforcement learning proposed by Qu Juntian's team at Tsinghua University, by integrating Long Short-Term Memory (LSTM) neural networks with fully connected neural networks, enables the deep reinforcement learning control strategy to consider both historical and current policy relationships, improving the control precision of the flexible arm's end position.

This flexible arm can not only perform delicate grasping tasks on land but also conduct operations like sampling in narrow, confined spaces underwater [16]. This research provides new ideas and methods for the intelligent application of cable-driven technology in complex underwater environments [16].

Cable-driven and hybrid actuation show promising application prospects in underwater operations. In detection and sampling tasks in complex underwater terrains, hybrid-driven flexible robotic arms can flexibly adjust actuation methods according to different operational needs, ensuring both motion precision and sufficient driving force and compliance, thereby improving operational efficiency and success rate [15].

2.4 Conclusion

The actuation methods for underwater soft hands and flexible robotic arms are continuously developing toward diversification and high performance. Pneumatic actuation is making constant breakthroughs in engineering applications, enhancing motion flexibility and control precision through structural and material innovations. Hydraulic actuation, aided by advanced sensors and intelligent algorithms, achieves high-precision control and plays a key role in heavy-load underwater operations. Cable-driven and hybrid actuation, through synergistic optimization, integrate the advantages of multiple actuation methods, providing more reliable solutions for complex underwater tasks.

3 Importance of Intelligent Control and Perception

When we talk about robots and intelligent systems, we often associate them with questions such as "Can they make their own decisions?" "Are they intelligent enough?" and "Can they understand the surrounding world?"

Imagine this scene: in a modern factory, a six-axis robotic arm is collaborating with a human operator. Unlike earlier generations of robots that "clumsily" repeated fixed motions, it now flexibly adjusts its grasping strategy according to changes in the shape and position of parts on the conveyor. When a worker reaches out to inspect a component, the robot, through vision and force feedback, recognizes that "a person is approaching" and proactively slows down to avoid danger. Underlying this behavior is the tacit synergy between intelligent perception and intelligent control. This example vividly shows that without intelligent perception, a robot is blind; without intelligent control, it merely stands still. Only when the two complement each other can a robot acquire the capability to "respond to complex scenarios as flexibly as a human."

Ultimately, these issues point to two core capabilities: Intelligent Control and Intelligent Perception. If we liken a robot to a human, control is its "brain and nervous system," determining how it forms strategies and executes motion; perception is its "sensory organs," helping it understand the external environment and its own state. With perception, it knows what is happening; with control, it can act appropriately on that information. Neither can be dispensed with.

In a certain sense, intelligent control and intelligent perception constitute the "soul" and the "reflexes" of modern robots and intelligent systems. They are not merely two technical modules, but the key that endows machines with intelligence. If earlier robots mechanically repeated preprogrammed routines, today's robots are moving toward "autonomy" and "adaptability"—no longer just expensive automated tools, but intelligent partners capable of companionship, collaboration, and even independent exploration.

To better understand these two concepts, we may first give brief definitions:

- Intelligent Perception refers to the process by which a robot or system acquires information via sensors (such as cameras, microphones, LiDAR, tactile sensors, etc.) and interprets it through algorithms. It enables the machine to "see," "understand speech," "sense touch," and even comprehend scenes and predict the future.
- Intelligent Control is the process by which the system, after processing perceptual information, takes actions and makes decisions. Relying on control theory and artificial intelligence methods, it ensures the robot not only moves but moves reasonably, smoothly, and efficiently.

These two components are mutually dependent: without perception, control is blind; without control, perception is static. A truly intelligent robot is necessarily the product of the deep integration of perception and control.

Why is their importance especially prominent today? Because the task environments faced by contemporary robots and intelligent systems are far more complex than in the past. Whether in flexible and changeable manufacturing workshops, congested urban roads, or delicate and precise surgical settings, robots must operate under highly dynamic, uncertain, and variable conditions. Traditional preprogrammed approaches often fall short because the real world is filled with continual change, requiring systems to have real-time responsiveness and adaptive adjustment. This is precisely where intelligent control and perception excel.

3.1 The Theoretical Foundations and Development of Intelligent Control

Intelligent control, as an important development direction in modern automation and intelligent systems, builds upon traditional control theory while integrating emerging technologies such as artificial intelligence and machine learning, thereby endowing control systems with autonomy, adaptability, and learning capability. Traditional control systems typically rely on accurate mathematical models and preset parameters and thus often struggle to achieve efficient and reliable control in complex, dynamic, and uncertain real environments. Intelligent control systems, however, can autonomously adjust control strategies through real-time perception of the environment and even continuously optimize performance through self-learning, achieving higher levels of intelligence and automation [17]. Core characteristics of intelligent control include: autonomy—making decisions without human intervention; adaptability—adjusting control parameters and strategies according to environmental changes; and learning capability—self-optimization and iteration via historical data and experience to improve control performance.

The evolution of intelligent control reflects a major shift from traditional control to intelligence-enhanced paradigms. Classical methods such as PID control and linear control are widely used in industrial automation, yet their heavy dependence on system models and parameters limits effectiveness in nonlinear, variable, and uncertain environments. With the continuous advancement of AI technologies and computational power, intelligent control has gradually become mainstream, traversing stages including fuzzy control, neural network control, imitation learning, and reinforcement learning. In 1965, Zadeh proposed fuzzy set theory, laying the foundation for fuzzy control. Fuzzy control does not rely on precise mathematical models; instead, it handles uncertainty and imprecise information through "if–then" rules and has been widely applied in household appliances and industrial process control [18]. With the explosive development of deep learning, neural networks have become the key driver of performance improvement in the field. Leveraging powerful nonlinear mapping and self-learning capability, they approximate the dynamic behavior of complex systems and automatically model input–output relationships through training data, making them suitable for high-dimensional and complex environments [19]. This also propelled the development of imitation learning, which quickly acquires effective control strategies by learning from expert or human demonstrations, while reinforcement learning optimizes behavioral strategies through interaction with the environment and reward mechanisms, significantly enhancing system autonomy and adaptability [17, 20].

As intelligent control theory has progressed, a range of advanced strategies has emerged. Fuzzy logic control suits scenarios where system models are unclear or environmental information is imprecise, achieving flexible decision-making via linguistic rules and fuzzy inference [18]. Reinforcement learning explores optimal strategies autonomously through reward and penalty signals and is now widely applied in robot navigation and manipulator control [17]. Deep learning-based control integrates deep neural networks to enable intelligent decision-making under high-dimensional perception and complex environmental constraints—deep reinforcement learning in particular shows great potential in autonomous driving and automated production [21].

Intelligent control technologies demonstrate remarkable advantages in nonlinear, variable, and uncertain environments. Traditional systems often exhibit insufficient robustness under environmental changes, whereas intelligent control can perceive the environment in real-time and dynamically adjust strategies, significantly improving stability and adaptability. For example, in autonomous vehicles, intelligent control flexibly adjusts driving behavior based on real-time road conditions, obstacles, and traffic signals, achieving safe and efficient autonomous driving [22]. Furthermore, intelligent control leverages multimodal data fusion, online learning, and adaptive optimization to cope with complex tasks and extreme operating conditions, meeting modern intelligent systems' requirements for high reliability and flexibility [20]. Ongoing advances have laid a solid theoretical and technical foundation for flexible manipulation and intelligent perception systems, driving intelligent systems toward higher levels of autonomy and intelligence.

3.2 *The Evolution and Role of Perception Technologies*

In robotics and intelligent systems, "perception" is typically defined as the process by which a robot uses sensors to collect external information and—through processing and interpretation—obtains an understanding of the environment, task, and its own state. This process is the starting point of robot intelligence and the prerequisite for control and decision-making. In other words, before a robot can decide "what to do," it must first "know what is happening."

In traditional automation scenarios, robots mostly operated in highly structured environments, such as automotive assembly lines. In such settings, the external environment is relatively static and controllable, so robots required minimal environmental perception capability. However, as applications expanded into unstructured, dynamic—even unpredictable—scenes such as autonomous driving, domestic services, surgical operations, or disaster rescue, environmental complexity surged. Reliance on preset motions or simplistic sensing became grossly insufficient. At this stage, the importance of perception technologies increased significantly: they directly determine whether the robot "sees clearly" and whether it can act reasonably as conditions change [23].

Common perceptual modalities—such as vision, audition, and touch—feed information into intelligent control systems. Visual sensors are often called the robot's "most important sensory organ." Vision supplies rich 2D and even 3D information for object detection, environmental structure recognition, and navigation. From early monocular and stereo systems for simple localization to today's deep learning-based visual understanding, vision has undergone rapid leaps. In practice, industrial robots rely on machine vision for part identification and quality inspection; autonomous vehicles use forward-facing cameras to recognize traffic signs, lane markings, and pedestrians; in medicine, surgical robots employ high-definition endoscopic cameras so surgeons can access microscale intraoperative detail in real-time [24]. Auditory sensing provides another important perceptual avenue. Microphone arrays and speech recognition systems have become standard in household service robots (e.g., smart speakers and voice assistants). They can not only capture speech content but also perform sound source localization for tracking. For instance, the Pepper robot uses speech perception and natural language processing to interact with humans, serving as a "bridge" in retail and domestic settings [25].

Tactile sensing primarily acquires pressure, texture, and force information at contact surfaces. Compared with vision and audition, touch provides more direct support for action execution. In industrial robotic grasping, tactile sensing prevents excessive force that could damage objects and preserves precision in assembly. In rehabilitation, exoskeleton robots equipped with force sensors can adjust assistance levels automatically when they sense that a patient's muscular output is weak, enabling personalized recovery [26].

Beyond these are higher-level modalities. LiDAR supplies accurate distance information over large areas, with its 3D point clouds widely used in environmental modeling and navigation. Autonomous vehicles and high-precision drones often treat LiDAR as a core sensing component. Waymo's autonomous vehicles construct real-time 3D maps through LiDAR and fuse them with visual data to achieve robust performance at night or in adverse weather. IMUs—comprising accelerometers, gyroscopes, and magnetometers—support posture estimation and motion tracking. Drones depend on IMUs for precise attitude information to maintain stable flight; wearable robots (intelligent prostheses) rely on IMUs to track user motion for natural, coordinated actuation.

These perception technologies supply information from different dimensions, complementing each other like multiple human senses to form a global understanding of the environment.

3.3 The Integration of Intelligent Control and Perception

3.3.1 Tight Coupling of Control and Perception: The Importance of Closed-Loop Systems

In robots and intelligent systems, perception and control are not independent segments; they are tightly coupled through closed loops. A closed loop refers to the process by which a robot acquires environmental and internal state information via perception, feeds that information into the control module for analysis and decision-making, then the execution results influence the environment, and new environmental changes are recaptured by the perception system. This perceive–control–act–perceive cycle constitutes the essential basis of autonomy [27]. Boston Dynamics' quadruped robot Spot is a typical illustration. During locomotion, if its visual sensors detect a change in step height, the control system immediately adjusts gait to maintain balance. Without perceptual feedback, Spot could not sustain dynamic stability over complex terrain (e.g., stairs, lawns, and gravel) [28]. Closed loops ensure not only motion stability but also safety in complex, uncertain environments.

The significance of closed-loop systems lies in transforming robots from "tools executing preset commands" into intelligent agents engaged in continuous dialogue with their environments, constantly refining behavior. This characteristic is especially critical in medical and service contexts. For example, the Da Vinci surgical robot relies on high-definition vision and force feedback, enabling surgeons to "control what they see"—a closed-loop interplay that ensures precision and safety.

3.3.2 Autonomous Behaviors Enabled by Perception-Guided Control Strategies

In complex environments, the choice of control strategy must be dynamically adjusted based on perceptual outcomes. Perception is not merely an "input signal"; it is the key that "guides action." For autonomous mobile robots (AMRs), LiDAR and vision jointly sense obstacle positions; the navigation system updates the map, and the control module adjusts trajectory accordingly. This "perception-driven control strategy" enables genuine autonomous navigation in unknown environments [23].

Autonomous vehicles represent the most intuitive application. While driving, a forward-facing camera detects a traffic light turning red; the perception system passes this to the control module, which issues a braking command. More complex still, when camera and radar detect a pedestrian crossing, the vehicle must not only stop but also decide—based on predictive outputs—whether to shift lanes for avoidance. Thus, perception guides both basic operations and higher-level behavioral patterns.

In industrial practice, the same mechanism appears frequently. A collaborative robotic arm uses vision to identify object positions and adjust grasping in real-time. If sensors detect a worker reaching in, the robot arm automatically adjusts its trajectory or pauses to avoid danger. This is essentially perception directly driving control strategy to achieve safe collaborative behavior [29].

3.3.3 Real-Time Decision-Making and Adaptive Capability Based on Perception

A robot's performance in dynamic, uncertain environments hinges on its capacity for real-time, perception-based decision-making and adaptation. Real-time capability means the system must swiftly convert sensor data into executable control commands—placing demanding constraints on perceptual bandwidth, computational efficiency, and controller response speed [17]. In UAV missions, IMUs provide attitude angles and acceleration data, while vision identifies obstacle positions. When a side gust is perceived, the control module must issue compensatory commands within milliseconds to correct trajectory; excessive delay could cause loss of stability and crash. This instantaneous responsiveness reflects the value of perception-based real-time decision-making.

Another example is intelligent rehabilitation robots. These systems use force and motion capture sensors to perceive patient states in real-time. When signs of muscular fatigue appear, the perception system flags the condition, and the control strategy adjusts assistance accordingly to prevent overexertion-related injury. This mechanism shows that real-time perception and decision-making not only enhance adaptability but also ensure safety and personalization in human–robot interaction [26].

3.3.4 Synergistic Enhancement Through Multimodal Perception and Intelligent Control

Single sensing modalities have limitations: cameras fail in darkness, LiDAR performance degrades in rain or snow, and IMUs can drift over long durations. Fusing heterogeneous sensors enables more robust multimodal perception, providing more accurate inputs for control and significantly elevating overall system intelligence [30]. Autonomous driving is the quintessential example: visual sensing supplies road and lane information; radar measures relative speed; LiDAR constructs spatial point clouds. Upon receiving fused data, the control system generates more optimized driving strategies. This multimodal control synergy empowers vehicles to handle illumination variation, adverse weather, and sudden events.

In human–robot collaboration, multimodality is equally vital. A collaborative robot identifies workpiece position via vision, senses grasping force through pressure sensors, and interprets spoken commands through speech recognition. The control system bases action decisions on these combined sources, enabling the robot to perform physical tasks while interacting naturally with humans [31]. Such fusion not only strengthens system robustness but also improves the naturalness and efficiency of interaction.

The integration of intelligent control and perception transforms robots from passive "actuators" into proactive "intelligent agents." Closed loops ensure robustness and safety; perception-driven control strategies enable genuine autonomy; real-time decisions allow rapid response; multimodal synergy further raises control accuracy and resilience.

3.4 The Practical Importance of Intelligent Control and Perception

With the rapid development of robotics and AI technologies, intelligent control and perception have permeated industry, household, transportation, and healthcare domains. This evolution not only reshapes production and lifestyle patterns but also poses new scientific challenges. Below, four representative application areas are discussed—alongside data processing, robustness, real-time performance, and reliability challenges—to highlight value, relevance, and future directions.

3.4.1 Applications in Industrial Automation: Intelligent Manufacturing and Warehouse Robotics

In modern industry, intelligent control and perception are reshaping manufacturing—shifting from “rigid assembly lines” to “flexible production.” In intelligent manufacturing, traditional robots rely on preprogrammed control and lack adaptive flexibility; with visual recognition, force control, and deep perception, robots now execute more complex autonomous tasks. For example, vision-guided assembly robots automatically identify part pose and location, and adaptive control algorithms dynamically adjust grasp strategies to avoid errors [32]. In intelligent inspection, robots detect defects early via multimodal perception, ensuring stable quality. In warehousing and logistics, Amazon’s Kiva robots are emblematic: hundreds of units traverse warehouse floors, navigating via QR codes and sensors, coupled with centrally scheduled dynamic control to perform flexible shelf handling. Compared with traditional methods, such systems multiply order processing speed and space utilization [33]. Kiva robots depend on closed-loop perception and control to avoid collisions, underscoring the importance of perception-driven control in large-scale cooperative systems.

3.4.2 Service and Domestic Robots: Autonomous Navigation and Interaction

As service robots enter homes and public spaces, the advantages of intelligent control and perception become more pronounced. Household environments are highly uncertain: furniture may shift, and human activity patterns are unpredictable. Thus, robots must integrate autonomous perception, real-time obstacle avoidance, and interactive communication. Early floor-cleaning robots from Ecovacs relied on simple bump sensors, leading to random, inefficient coverage. Modern products feature LiDAR and cameras to build indoor maps and optimize paths; intelligent control strategies ensure maximal coverage under limited battery capacity. This leap from “blind wandering” to “purposeful operation” stems from deep perception–control integration [32]. In service contexts, human–robot interaction is also critical. The Pepper social robot uses visual recognition and speech perception to converse naturally. Combined with control systems, Pepper modulates interaction style based on user emotion and tone [25]. Such robots not only perform physical tasks but also provide social companionship, highlighting the role of perception and control in “understanding humans.”

3.4.3 Autonomous Transportation Systems: Driverless Vehicles and Intelligent Traffic Management

Intelligent transportation is a canonical setting for integrating perception and control, most vividly demonstrated in autonomous driving. Autonomous vehicles rely on cameras, LiDAR, millimeter-wave radar, and IMUs to build real-time environmental maps; control modules must parse semantic context and generate safe, efficient driving strategies. Waymo's driverless taxi services have begun pilot operations in parts of U.S. cities. In deployment, Waymo leverages 360° LiDAR for surrounding environment sensing, neural networks for scene understanding, and probabilistic control models for motion selection, enabling safe operation in highly dynamic urban traffic [34].

Future intelligent transportation systems will additionally emphasize centralized traffic optimization. In vehicle–infrastructure cooperative trials, roadside sensors and vehicles share information to achieve global coordination. Through edge computing and low-latency communication, vehicles obtain traffic signal states and neighboring vehicle speeds, enabling more intelligent regulation to reduce congestion and accidents.

3.4.4 Medical Robotics: Surgical and Rehabilitation Assistance

Healthcare is among the application domains most demanding of high precision and reliability in perception–control integration. In surgical robotics, high-definition vision and tactile feedback extend the surgeon's capabilities. The Da Vinci system offers more than tenfold magnified 3D imaging and, coupled with intelligent controlled arms, enables remote operation and minimally invasive incisions. Clinical evidence shows its application in urological and cardiac procedures significantly shorten patient recovery time [35].

In rehabilitation, exoskeleton robots employ IMUs and force sensors to monitor patient motion and adjust assistance intensity in real-time. For stroke rehabilitation, if muscular weakness is sensed, the control system amplifies support to ensure therapeutic effect while preventing injury [26]. These applications show that perception–control synergy is not merely a technical upgrade, but essential to safety and personalization in medical contexts.

3.4.5 Challenges and Solutions: Data Processing, Robustness, Real-Time Performance, and Reliability

Despite success across multiple scenarios, several challenges remain:

Data Processing Complexity: Modern robots process massive multisource sensor streams. For instance, autonomous vehicles generate several gigabytes of perceptual data per second, stressing bandwidth and computation. Edge computing and distributed processing help mitigate latency and bandwidth bottlenecks [36].

Insufficient Robustness: Sensors are sensitive to environmental factors—strong light impacts cameras, adverse weather degrades LiDAR. Future directions involve stronger sensor fusion and robust algorithms to maintain stability under adverse conditions [30].

Real-Time Requirements: Many scenarios demand millisecond-level response (e.g., UAV obstacle avoidance and emergency braking in autonomous driving). Solutions include hardware acceleration (GPU/FPGA), lightweight deep models, and 5G/6G ultra-low-latency communication.

Reliability and Safety: In surgical robotics and autonomous driving, reliability directly concerns human life. Current approaches emphasize redundancy and adaptive fault tolerance to ensure baseline functionality even if some sensors or control units fail [37].

In sum, intelligent control and perception have already demonstrated substantial value in industrial manufacturing, domestic services, intelligent transportation, and healthcare. They elevate efficiency and service quality and accelerate the intelligent upgrading of transportation and medicine. However, large-scale, stable societal deployment still requires solutions to data processing, robustness, real-time performance, and reliability challenges. With advances in sensors, computational architectures, and AI algorithms, these barriers are being progressively reduced. In the future, intelligent perception and control will become core pillars benefitting human society at scale.

3.5 *Future Trends and Challenges*

Over the past decades, intelligent control and perception technologies have gradually become two central pillars of intelligent systems. From industrial automation to service robots, from autonomous driving to surgical assistance, their deep integration not only advances technology but also subtly reshapes societal operations. Yet development will not plateau at current levels. With rapid progress in AI, big data, and network infrastructure, intelligent control and perception are entering a new stage.

3.5.1 Future Technologies in Intelligent Control and Perception

Deep Integration of AI and Intelligent Control: Traditional control's reliance on models constrains adaptability in complex environments. With deep learning and reinforcement learning, future intelligent control will gain "adaptive evolution" capability [17]. Controllers will not only execute preset rules but also optimize through accumulated experience and data feedback. Autonomous driving programs at Tesla and Waymo have begun introducing deep reinforcement learning so vehicles can learn strategies via trial-and-error in simulation. This technology may extend to domestic service robots, enabling personalized behavioral patterns through prolonged interaction with household members.

The Rise of Edge Computing: Processing massive sensor data streams has long been a bottleneck. Edge computing and distributed intelligence will become key solutions [36]. By processing data directly at the device, robots can complete decisions within milliseconds rather than waiting on the cloud. In smart factories, collaborative robots can compute real-time sensing locally to rapidly correct assembly errors; in smart transportation, vehicles leverage roadside edge nodes to access traffic state promptly and adjust control behavior.

Further Deepening of Multisensor Fusion: Future robots will depend on richer multimodal perception, achieving fusion from low-level physical signals to high-level semantic information. The fragility of single sensors will be mitigated through multisource fusion. For example, future UAVs may combine vision, infrared thermal imaging, and acoustic sensing together with meteorological data for more stable flight control. In medical robotics, multisensor fusion will allow clearer delineation of vessels, nerves, and tumors to support real-time minimally invasive procedures [38]. This trend toward fused perception and control will move robots closer to human-like multisensory integration.

3.5.2 Potential Impact on Future Intelligent Systems

Comprehensive Autonomy: As perception and control achieve deep synergy under AI and edge computing, autonomy will increase across domains. Autonomous vehicles will require fewer human interventions; factory production lines will reconfigure processes based on real-time sensing; household robots will evolve from "cleaning devices" to "life assistants," proactively caring for elderly residents. This shift transforms machines into "intelligent partners" rather than passive tools [37].

Ubiquitous Adoption and Large-Scale Deployment: Future trends emphasize not only performance breakthroughs but widespread adoption. With declining sensor costs and more energy-efficient chips, intelligent control and perception will expand from high-end industries to everyday life. Low-cost vision and tactile modules will let SMEs adopt intelligent manufacturing; home assistance robots may become as commonplace as smartphones. In this trajectory, intelligent control and perception will become foundational societal technologies akin to electricity or the Internet.

Building an Intelligent Society: In the long term, intelligent control and perception will become the cornerstone of an intelligent society. Coordinated transportation reduces accidents and congestion; medical intelligent devices enhance health assurance; service robots in public spaces mitigate labor shortages. Human–machine collaboration will intensify, yielding leapfrog improvements in societal operational efficiency [37].

3.5.3 Challenges and Future Directions

Despite promising prospects, multiple challenges must be confronted:

- Data and Privacy: Robots process large volumes of personal or proprietary data; privacy protection and secure transmission are critical. Federated learning and other privacy-preserving AI methods will be required.
- Robustness and Reliability: Sensors are vulnerable to noise, weather, and communication disruptions; more resilient fusion algorithms and adaptive mechanisms are needed.
- Real-Time Performance and Scalability: As deployments scale, achieving rapid responses under constrained resources remains difficult. Lightweight AI models and low-latency (e.g., 6G) communication technologies will be key.
- Ethics and Societal Impact: High autonomy raise accountability and ethical concerns—such as liability in autonomous driving accidents or regulatory boundaries for medical robot decision scope.

Future research should focus on "AI-driven adaptive control," "fusion of multimodal perception with edge intelligence," and "high-reliability system architectures." Meanwhile, policy and ethical frameworks must advance in parallel to ensure beneficial societal impact and mitigate emerging risks.

Intelligent control and perception are the dual engines propelling the continual evolution of robots and intelligent systems. Future trends will manifest in three dimensions: deep fusion with AI, the proliferation of edge computing, and coordinated multimodal sensing and control. These advances will drive robots toward genuine autonomy and large-scale adoption, facilitating the formation of an intelligent society. Facing the future, however, we must address challenges in data privacy, robustness, real-time performance, and socio-ethical governance. Through strengthened algorithmic research, enhanced hardware infrastructure, and the establishment of legal and ethical standards, humanity will better steer this technological transformation. Intelligent control and perception are not only the "senses and brain" of intelligent systems, but also the core driving force behind transformative change in industry, society, and human–machine relations—ushering in a new stage of collaboration and symbiosis.

References

1. Qu, J., Zhang, X., Wang, Y., et al. (2024). A cable-actuated soft manipulator for dexterous grasping based on deep reinforcement learning. *Advanced Intelligent Systems, 6*(10), 2400119.
2. Mingxin, W., Afridi, W. H., Jiaxi, W., Afridi, R. H., Wang, K., Zheng, X., Wang, C., & Xie, G. (2024). Octopus-inspired underwater soft robotic gripper with crawling and swimming capabilities. *Research, 7*, 0456.
3. Liu, S., Mei, T., Xu, W., Meng, F., Jiao, J., & Su, H. (2024). Research status and development trends of flexible robotic arms. *Robot Technique and Application, 217*(1), 7–17.
4. Advances in flexible robotic manipulator systems—Part I: Overview and dynamics modeling methods (2024). IEEE .
5. Xie, Z., et al. (2023). Octopus-inspired sensorized soft arm for environmental interaction. *Science Robotics, 8*, eadh7852.
6. Verheyen, C. A., Uzel, S. G., Kurum, A., Roche, E. T., & Lewis, J. A. (2023). Integrated data-driven modeling and experimental optimization of granular hydrogel matrices. *Matter, 6*(3), 1015–1036.
7. Yan, J., Shi, P., Zhang, X., et al. (2018). Review on development of bionic mechanism, drive, modeling and control for soft robotic arms. *Journal of Mechanical Engineering, 54*(15), 1–14.
8. Nilsson, N. J. (2009). "Hand-eye" research. In *The quest for artificial intelligence* (pp. 141–148). Cambridge University Press.
9. Song, X. (2002). Electroactive polymer: The preferred material for simulating muscles. *Foreign Science and Technology Trends, 6*, 18–22.
10. Yan, J. (2023). *Grasping mechanical modeling and simulation of pneumatic soft robotic hand.* Harbin Institute of Technology.
11. Laschi, C., Cianchetti, M., Mazzolai, B., et al. (2012). Soft robot arm inspired by the octopus. *Advanced Robotics, 26*(7), 709–727.
12. Li, Y., Goh, G. C., Wang, Z., et al. (2024). Disturbance-adaptive tapered soft manipulator with precise motion controller for enhanced task performance. *IEEE Transactions on Robotics, PP*(99), 1–20.
13. Marchese, A. D., Katzschmann, R. K., & Rus, D. (2015). Design, kinematics, and control of a soft spatial fluidic elastomer manipulator. *International Journal of Robotics Research, 35*(7), 840–869.
14. Kobayashi, W., Ito, K., & Yamamoto, S. I. (2014). Displacement control of water hydraulic McKibben muscles with load compensation. *JFPS International Journal of Fluid Power Systems, 8*(2), 107–112.
15. Shi, H., Tan, K., Zhang, B., et al. (2022). Review on research progress of hydraulic powered soft actuators. *Energies, 15*(23), 9048..
16. Wei, H., Liang, C., Wang, J., et al. (2024). Research progress of cable-driven flexible manipulators and their application in agricultural robots. *Journal of Intelligent Agricultural Equipment (Chinese and English), 5*(4), 95–106.
17. Sutton, R. S., & Barto, A. G. (2018). *Reinforcement learning: An introduction* (2nd ed.). MIT Press.
18. Zadeh, L. A. (1965). Fuzzy sets. *Infection Control, 8*(3), 338–353. https://doi.org/10.1016/S0019-9958(65)90241-X
19. Haykin, S. (1999). *Neural networks: A comprehensive foundation* (2nd ed.). Prentice Hall.
20. Kober, J., Bagnell, J. A., & Peters, J. (2013). Reinforcement learning in robotics: A survey. *The International Journal of Robotics Research, 32*(11), 1238–1274. https://doi.org/10.1177/0278364913495721
21. LeCun, Y., Bengio, Y., & Hinton, G. (2015). Deep learning. *Nature, 521*(7553), 436–444. https://doi.org/10.1038/nature14539
22. Chen, L., Li, Y., Huang, C., et al. (2022). Milestones in autonomous driving and intelligent vehicles: Survey of surveys. *IEEE Transactions on Intelligent Vehicles, 8*(2), 1046–1056.
23. Thrun, S. (2002). Probabilistic robotics. *Communications of the ACM, 45*(3), 52–57.

24. Surgical I. da Vinci. (2013). Surgical system. http://www.Intusurg.com/html/davinci.html
25. Pandey, A. K., & Gelin, R. (2018). A mass-produced sociable humanoid robot: Pepper: The first machine of its kind. *IEEE Robotics & Automation Magazine, 25*(3), 40–48.
26. Dollar, A. M., & Herr, H. (2008). Lower extremity exoskeletons and active orthoses: Challenges and state-of-the-art. *IEEE Transactions on Robotics, 24*(1), 144–158. https://doi.org/10.1109/TRO.2008.915453
27. Brooks, R. A. (1991). Intelligence without representation. *Artificial intelligence, 47*(1–3), 139–159. https://doi.org/10.1016/0004-3702(91)90053-M
28. Raibert, M. H. (1990). Trotting, pacing and bounding by a quadruped robot. *Journal of Biomechanics, 23*, 79–98.
29. Matheson, E., Minto, R., Zampieri, E. G., Faccio, M., & Rosati, G. (2019). Human–robot collaboration in manufacturing applications: A review. *Robotics and Computer-Integrated Manufacturing, 58*, 100744. https://doi.org/10.1016/j.rcim.2019.100744
30. Wang, Z., Wu, Y., & Niu, Q. (2019). Multi-sensor fusion in automated driving: A survey. *IEEE Access, 8*, 2847–2868.
31. Goodrich, M. A., & Schultz, A. C. (2007). Human–robot interaction: A survey. *Foundations and Trends in Human–Computer Interaction, 1*(3), 203–275. https://doi.org/10.1561/1100000005
32. Zhang, B., Wang, J., Rossano, G., et al. (2011). Vision-guided robotic assembly using uncalibrated vision. In *2011 IEEE International Conference on Mechatronics and Automation* (pp. 1384–1389). IEEE.
33. Wurman, P. R., D'Andrea, R., & Mountz, M. (2008). Coordinating hundreds of cooperative, autonomous vehicles in warehouses. *AI Magazine, 29*(1), 9–20. https://doi.org/10.1609/aimag.v29i1.2082
34. Schwall, M., Daniel, T., Victor, T., et al. (2020). Waymo public road safety performance data. arXiv preprint arXiv:2011.00038.
35. Lanfranco, A. R., Castellanos, A. E., Desai, J. P., & Meyers, W. C. (2004). Robotic surgery: A current perspective. *Annals of Surgery, 239*(1), 14–21. https://doi.org/10.1097/01.sla.0000103020.19595.7d
36. Satyanarayanan, M. (2017). The emergence of edge computing. *Computer, 50*(1), 30–39. https://doi.org/10.1109/MC.2017.9
37. Jordan, M. I., & Mitchell, T. M. (2015). Machine learning: Trends, perspectives, and prospects. *Science, 349*(6245), 255–260. https://doi.org/10.1126/science.aaa8415
38. Makridakis, S. (2017). The forthcoming artificial intelligence (AI) revolution: Its impact on society and firms. *Futures, 90*, 46–60. https://doi.org/10.1016/j.futures.2017.03.006

Structure and Materials of Flexible Manipulators

1 Classification of Flexible Manipulators

As an important branch of robotics, soft manipulators have demonstrated significant potential in recent years across industrial automation, medical rehabilitation, and service robotics [1]. Compared with traditional rigid manipulators, soft manipulators offer greater adaptability and safety, enabling more effective handling of grasping tasks in uncertain environments. Gripper-type soft manipulators emulate biological grasping mechanisms to achieve stable handling of irregular and fragile objects, thereby overcoming the limitations of conventional rigid grippers [2].

1.1 Classification of Soft Actuators Based on Actuation Principles

Pneumatic Soft Grippers: Pneumatically actuated soft grippers are among the earliest developed soft grasping technologies. They achieve grasping actions by controlling pneumatic or hydraulic pressure. These grippers typically employ elastic materials (such as silicone rubber) to form actuators that generate bending deformation through inflation and deflation of internal chambers. The working principle is based on the PneuNet (pneumatic network) structure: when compressed air is injected into internal chambers, the asymmetry of the material structure causes the chambers to expand and bend in a specific direction, thereby realizing grasping functions [2]. This mode of grasping can adaptively conform to objects that are irregular, fragile, or of unknown size, greatly expanding the application boundaries of robots in industrial automation, medical rehabilitation, agricultural harvesting, and scientific exploration.

J. Qu, *Robotic Flexible Manipulation and Intelligent Sensing Technologies*,
https://doi.org/10.1007/978-981-95-6232-9_2

Representative examples include three-finger pneumatic grippers, in which each finger contains multiple independent air chambers. By precisely controlling the pressure in each chamber, the gripper can implement various grasping modes such as enveloping and pinching. Advantages include structural simplicity, low cost, and gentle interaction with objects, making such grippers particularly suitable for handling fragile items like food and electronic products.

1.1.1 Core Actuation Principle

Pneumatic soft actuators (PSAs) constitute the fundamental units of soft grippers [3]. The most basic operating model can be summarized as a sealed air chamber made of hyperelastic materials (e.g., silicone rubber) and one or more constraint layers. When compressed air is injected into the chamber, the internal pressure rises, and the chamber tends to expand in all directions. However, the constraint layer—typically made from materials that are inextensible or have far lower extensibility than the primary material (such as fabrics, fibers, or thin sheets of harder polymers)—limits deformation in specific directions. This "asymmetric expansion" ultimately induces directional, predictable motions in the actuator, such as bending, elongation, or twisting.

In a typical bending actuator [4], the base usually integrates an inextensible constraint layer. Upon inflation, the elastic material at the top and sides of the actuator expands, while the base cannot elongate due to the constraint layer. The resulting difference in length causes the entire actuator to bend toward the side of the constraint layer. By precisely regulating the input air pressure, the bending angle of the actuator can be continuously adjusted, thereby enabling fine-grained control of grasping.

1.1.2 Key Components

A complete pneumatic soft gripper system typically comprises the following components: (1) Soft actuation unit: The primary element executing the grasping action, usually consisting of one or more soft fingers. These fingers are fabricated from hyperelastic materials such as silicone rubber using manufacturing processes, including mold casting or 3D printing. (2) Pneumatic control system: Including the air supply (e.g., air compressor or pump), solenoid valves, proportional valves, pressure regulators, and connecting pneumatic lines. This system precisely controls the direction, flow rate, and pressure of the air delivered to each soft actuation unit and is critical for achieving accurate gripper motion. (3) Sensing and feedback system (optional): To enable closed-loop control and more intelligent grasping, various sensors may be integrated into the soft gripper—for example, flexible bending sensors, pressure sensors, or tactile sensors—to monitor in real-time the gripper's morphology and contact forces with objects. (4) Structural support and connectors: A base used to mount and arrange multiple soft fingers into specific configurations (e.g., two-finger, three-finger, or anthropomorphic five-finger), as well as a flange for attaching the gripper to the robot arm's end-effector.

The PneuNet gripper developed at Harvard University [5] can achieve coordinated multi-finger grasping via a single pneumatic input. Its advantages include high output force and structural simplicity; however, it requires an external air supply and exhibits relatively slow response.

After years of development, researchers have proposed a variety of pneumatic soft gripper designs based on different physical principles and structural innovations, each with distinct forms and functions. The following are several representative designs.

PneuNets (Pneumatic Networks) Actuators: PneuNets are a widely used bending actuator structure first introduced by the Whitesides group at Harvard University [6]. Internally, they feature a series of small air chambers connected in series via narrow channels. A sheet of inextensible paper or fabric is embedded at the base as a strain-limiting layer. Upon inflation, internal pressure causes the chambers to expand like an accordion bellows. Due to the presence of the bottom constraint layer, expansion is concentrated in the top and side walls of the actuator, producing smooth bending motion with large curvature. By adjusting chamber geometry, wall thickness, and the position of the constraint layer, the bending characteristics of the actuator can be tailored. Multiple PneuNets actuators can be combined to form efficient soft fingers and grippers.

Fiber-Reinforced Actuators: Fiber-reinforced actuators constitute another classical soft actuator design [7, 8]. They typically consist of an elastic air chamber with a cylindrical or rectangular cross-section, externally wound or embedded with inextensible fibers (e.g., Kevlar threads) in specific patterns. These fibers function analogously to connective tissues in human muscle, precisely constraining the direction of chamber expansion. For example, helical fiber windings can produce simultaneous bending and twisting upon inflation; crossed ring windings can effectively prevent radial "bulging," promoting primarily axial elongation or bending. Fiber-reinforced actuators offer advantages such as structural simplicity, ease of fabrication, and strong pressure-bearing capability and are widely applied in grasping tasks that require relatively large output forces.

Granular Jamming Grippers: The design of granular jamming grippers is inspired by a physical phenomenon known as the "jamming transition" [9]. Such grippers typically employ a soft bladder filled with granular materials (e.g., coffee grounds or glass microspheres). In the unactuated state, the particles are loosely packed and the bladder is highly compliant, allowing it to conform to the surface of objects much like a fluid. Once good contact with the object is established, a vacuum pump evacuates air from the bladder; external atmospheric pressure compresses the particles together. The resulting high static friction among particles drives a rapid transition of the system from "fluid-like" to "solid-like," thereby locking the object securely in place. The principal advantage of this design lies in its exceptional generality: in principle, it can grasp objects of arbitrary shape without requiring complex sensors or control algorithms.

1.2 SMA-Based Grippers

The shape memory effect refers to the phenomenon whereby an SMA material, after undergoing plastic deformation in the low-temperature martensitic state, can fully recover its deformation and return to a preset original shape once heated above a specific transformation temperature (the austenite finish temperature, Af), corresponding to the high-temperature austenitic phase. This process is one-way: upon cooling, the material retains the shape it had at high temperature unless external force is applied to deform it again. In soft grippers, SMAs are typically fabricated as wires, springs, or thin films and serve as the primary actuation elements. Their actuation mechanism principally exploits the shape memory effect [10, 11], comprising the following steps: (1) Programming: In the high-temperature austenitic state, the SMA is set to the desired "original" shape (e.g., the gripper in an open configuration). (2) Deformation: After cooling to the low-temperature martensitic state, external force is applied to deform the SMA (and the connected soft structure) to another shape (e.g., the gripper in a closed configuration). (3) Activation: The SMA is heated above the austenite transformation temperature (commonly via electrical Joule heating). Driven by the shape memory effect, the SMA contracts or bends, striving to restore its "original" shape, thereby actuating the soft structure to perform grasping or release. (4) Cooling/Reset: After heating ceases, the SMA cools and retains the high-temperature shape. To return to the deformed state, external restoring forces are typically required (e.g., elastic elements, antagonistic SMA actuators, or the load itself).

The most prevalent activation method is electrical Joule heating, wherein current is passed through the SMA and resistive heating triggers the phase transformation.

A typical SMA-based soft gripper generally consists of the following components [12]: SMA actuator: Typically a wire or spring made of nickel–titanium (NiTi) alloy, serving as the core force-generating element. Soft matrix: Acting as the gripper's "skeleton" and "muscle," usually fabricated from polymeric elastomers such as silicone or polyurethane, providing compliant deformation capability. Constraint/transmission mechanism: Converts the SMA's linear contraction into bending motion of the gripper fingers. Power and control system: Supplies electrical energy to the SMA and precisely regulates heating duration and power, thereby modulating the gripper's motion. For example, the SMA tendon-driven actuator developed at Sungkyunkwan University [12] employs a long-tendon design to transform linear contraction into bending, achieving a maximum bending angle of up to 400° and a gripping force of 30 N. Advantages include high power density and silent operation; however, cooling is slow and energy efficiency is relatively low.

1.2.1 Typical Structural Designs

Tendon-Driven Mechanism: This design emulates biological tendon systems, arranging SMA wires as "artificial tendons" within channels located inside or on the exterior of soft fingers. When the SMA wire is heated and contracts, it pulls the soft finger like a tendon, inducing bending. Studies [11, 13] demonstrate that introducing SMA wires as freely sliding tendons within a polymer matrix can significantly enhance bending angle and output force. Advantages include replaceable SMA wires and high design flexibility: by modifying the tendon routing path, complex deformation modes (e.g., twisting or S-shaped bending) can be achieved.

Variable-Stiffness Mechanism: A persistent challenge for conventional SMA soft grippers is their inherent compliance, which limits gripping force. To address this, researchers have proposed variable-stiffness concepts. By integrating a second smart or phase-change material into the soft structure, the gripper's stiffness can be significantly increased on demand, enabling higher gripping forces while maintaining compliance. A common approach combines SMA actuators with low-melting-point materials (e.g., paraffin wax) or shape memory polymers (SMPs). Prior to grasping, the gripper remains soft and can conform to the object. After grasping, heating or cooling solidifies the wax or hardens the SMP, thereby "locking" the gripper's shape, markedly increasing stiffness and enabling secure holding of heavy objects [14].

1.3 Dielectric Elastomer Grippers

Dielectric elastomer (DE) grippers deform under electric fields due to Maxwell stress. Their structure typically comprises three layers: an elastomer film sandwiched between two compliant electrodes. When energized, the film compresses in thickness and expands in-plane [15, 16]. For example, hydraulically amplified self-healing electrostatic (HASEL) grippers combine electrostatic and hydraulic principles, achieving muscle-like contraction with self-healing capability. Advantages include rapid response and large strain; however, they require high-voltage power supplies (on the order of several kilovolts) and are susceptible to electrical breakdown. Dielectric elastomer actuators (DEAs) are structurally simple, with a canonical "sandwich" configuration [15]: a compliant dielectric elastomer film is clamped between two compliant electrodes. Dielectric elastomer layer: The core of the actuator, typically made from polymers with high dielectric constant, low elastic modulus, and high breakdown strength. The most widely studied materials include acrylics (e.g., VHB-series tapes) and silicones. These materials impart excellent stretchability and flexibility. Compliant electrode layers: The electrodes must remain conductive under large deformation of the elastomer. Common compliant electrode materials include carbon-based media (e.g., carbon powder and carbon nanotubes) and conductive polymers. These are typically applied to the elastomer surface by coating or printing. The actuation mechanism of DEAs is based on the physical phenomenon known as Maxwell stress. The operation can be delineated as follows: Voltage application: When a high voltage (typically at the kilovolt level) is applied between the upper and lower electrodes, opposite charges accumulate on the electrodes, creating a strong electric field within the dielectric elastomer film. Electrostatic force generation: Oppositely charged electrodes attract, producing a strong electrostatic force between the electrode layers. This force compresses the elastomer in the thickness direction. Material deformation: Because dielectric elastomers are approximately incompressible (volume remains nearly constant), compression in thickness necessarily induces expansion and stretching in the planar directions. Completion of actuation: When the voltage is removed, the electric field disappears, the Maxwell stress vanishes, and the elastomer film rapidly returns to its initial state driven by its elastic restoring force.

1.3.1 Typical Gripper Structure Design

Based on the above principles, researchers have designed a variety of dielectric elastomer flexible grippers with different shapes and unique functions [16]. Finger-like Grippers: This is the most common design, imitating the structure of the human hand. By arranging two, three, or more unidirectional bending drivers side by side as "fingers," enveloping grasping of objects can be achieved. This design has a simple structure and intuitive control and is suitable for grasping objects of various shapes.

Spring-Roll Grippers: This design takes advantage of the elastic potential energy of pre-stretched film. The pre-stretched DEA is combined with a flexible frame. When there is no voltage, the film is in a curled state. After applying voltage, the film stretches and flattens, and the fingers "unfold." This design can achieve very fast grasping actions.

Cup-Shaped/Adaptive Grippers: This type of gripper is typically composed of a circular DEA driver. When activated, the edge of the driver shrinks toward the center, forming a cup-shaped structure that can gently wrap around irregular or fragile objects like a bag, demonstrating excellent adaptability.

Conical DEA Grippers: By using conical dielectric elastomer actuators, multi-degree-of-freedom coupled movements such as stretching and bending can be achieved, thereby constructing more powerful multifunctional grippers that can complete more complex control tasks.

2 Structural Design of Flexible Manipulators

2.1 Design Principles of Flexible Manipulation Structures

Flexible manipulation structures—particularly in soft and continuum robotics—are grounded in a set of interrelated design principles that together enable systems to safely, robustly, and effectively interact with unstructured environments. The following subsections examine these principles: Compliance and adaptability; modularity and reconfigurability; integration of sensing and actuation; lightweight and compactness; and bio-inspiration in structural design.

Compliance and Adaptability: Compliance refers to the capacity of a manipulator to undergo elastic deformation under load and return to its original shape. This intrinsic softness enhances safe interaction with humans and unstructured objects, as well as robustness in uncertain environments. In classical rigid robotics, stiffness is often desirable—but in flexible manipulators, compliance enables passive force absorption, shape conformation, and tolerance to positioning errors [16]. Soft robotics literature emphasizes adaptability through compliance. For example, continuum manipulators—such as octopus-inspired arms—exploit large-deformation capacities and high degrees of freedom to conform passively to surfaces, facilitating grasping of unknown or irregularly shaped objects. Pneumatic structures and elastomeric bodies absorb impacts and adjust their shape dynamically, which enhances both interaction safety and task generality [17].

Modularity and Reconfigurability. Modularity divides a manipulator into discrete, interacting units or segments—each capable of specialized manipulation roles. Reconfigurability enables these modules to rearrange, replace, or retrain their connections to match new tasks, environments, or damage scenarios. In soft and continuum robotics, this concept is advanced via modular systems such as SoftSnap—snap-together modules incorporating actuation, sensing, and control, enabling rapid system reconfiguration into a range of morphologies (e.g., starfish-like, snake-like) [18]. Similarly, self-reconfiguring modular robots use homogeneous modules with standard interfaces to adapt morphology for novelty tasks or structural repair. Modularity enhances scalability, maintainability, and functional versatility. Using a common set of soft-elastomeric or hybrid modules reduces manufacturing cost and accelerates field adaptation.

Integration of Sensing and Actuation: To deliver truly flexible manipulation, sensing and actuation must be co-designed and embedded within the structure. Distributed, soft sensors provide proprioceptive and tactile feedback, while embedded actuators actuate shape change. Recent advances include coaxially printed core-sheath fibers combining actuation and sensing and self-sensing tensile valves for untethered soft systems [19]. Capacitive or piezoresistive soft sensors can measure deformation and contact forces, enabling closed-loop control without rigid external sensors. Examples such as SoPrA (Soft continuum

Proprioceptive Arm) integrate internal flex sensors within fluidic actuators, enabling real-time modeling and force estimation without external instrumentation [20]. Integration supports embodied intelligence: The mechanical structure and sensor layout simplify control by exploiting passive environmental interactions.

Lightweight and Compactness Considerations. Lightweight and compact structural design is vital for robotic mobility, safety in human-adjacent settings, and deployable systems (e.g., wearables and space platforms). Soft materials—including silicone, elastomers, and hydrogels—are inherently low density, facilitating compact systems with high workspace-to-weight ratios. Origami- and kirigami-based mechanisms further enhance compactness via folding strategies: the Yoshimura-ori pattern, combined with shape memory alloys (SMA), yields earthworm-like actuators capable of 3D deformation and deployment while occupying minimal stowed volume. Tensegrity structures—networks of cables and compressive elements—provide low-mass, high-strength compliance, valuable in space exploration (e.g., NASA's Super Ball Bot) and soft modular robots.

Bio-inspiration in Structural Design: Nature is the preeminent source of compliant, adaptable, and efficient manipulation structures. Iconic examples include octopus arms, elephant trunks, caterpillar segments, and vine growth—each illustrating extreme adaptability, multi-DOF motion, and functional pruning. Bio-inspired design yields strategies such as continuum soft manipulators (modeling octopus) and peristaltic vine-like extensions (e.g., thin-pressurized growth soft robots) [21]. Tensegrity architectures replicate cytoskeleton-like tension networks observed in cells—providing dynamic compliance and load redistribution. In articulated soft robotics, hybrid designs combine rigid bones and soft muscle-like actuators, mimicking vertebrate musculoskeletal systems to balance structural support with compliance.

The above principles coalesce into a unifying design philosophy: Employ compliant materials and structures to safely engage with environments; embed sensors and actuators for real-time feedback and control; modularize for flexibility and scalability; design for minimal mass and stowage; and draw from biological systems that have evolved superior morphological intelligence.

Future work spans AI-driven topology optimization of compliant structures, integration of self-healing and variable-stiffness materials, and holistic co-design of structure, sensing, control, and learning. This integrated, bio-inspired framework lays the groundwork for flexible manipulation architectures that are robust, adaptive, and intelligent.

2.2 *Structural Configurations and Mechanisms*

The design of flexible manipulation structures encompasses a diversity of mechanisms that enable compliant, adaptive, and dexterous interaction. This section examines five principal configurations: tendon-driven architectures; pneumatic and fluidic systems; origami- and kirigami-based designs; soft continuum manipulators; and hybrid rigid–soft constructs. Each approach offers distinctive capabilities and presents unique design trade-offs.

Tendon-Driven Structures: Tendon-driven configurations emulate musculoskeletal systems by routing cables or tendons through a flexible backbone or joint chain. Actuation is effected by tensioning one or more tendons, producing bending and motion. Such systems are prized for their positional precision, compact actuation, and controllable stiffness. Early continuum robots like the tensor-arm and COAST guidewire robot employ tendon routing through nested links or disks to enable curvature control in tight, constrained environments [22–24]. The soft self-pumping actuator (SSPA) further integrates tendons to actuate two pneumatic chambers, effectively combining pneumatic scalability with tendon precision and enhanced energy efficiency [22]. The BPI SoftHand uses underactuated tendon routing to realize complex grasp synergies with a single motor, leveraging elastic compliance while preserving robust object conformity [25]. Recent developments in tendon-driven continuum manipulators incorporate ball-jointed disk chains actuated by tri-tendon arrangements, enabling multi-directional bending with simplified mechanical design. Tendon-driven systems can also incorporate variable-stiffness mechanisms, as seen in modular manipulators that adjust tension and internal routing to improve payload capacity and tip stability.

Pneumatic and Fluidic Architectures: Pneumatic actuation harnesses pressurized fluid within elastomer chambers to generate bending, extension, or twisting motions. Variants include fiber-reinforced muscles, vacuum-actuated structures, and McKibben-style braided actuators. Their advantages lie in smooth, large-range deformation, inherent compliance, and safe environmental interaction. Kirigami-enhanced skins on pneumatic actuators transform simple inflation into sophisticated shape changes, such as crawling motion or twist-and-extend dynamics. These skins, formed by patterned cuts on planar sheets, buckled into pop-up 3D geometries under internal pressure. Their performance is highly tunable via cut geometry, and fabrication via laser or plotter cutting offers scalability across various materials and size scales [26]. In systems that combine tendons with pneumatic components, tensioned tendons modulate the pneumatic chamber deformation, enabling variable stiffness and precise actuation—as demonstrated in tendon-driven pneumatic grippers [27].

Origami- and Kirigami-Based Configurations. Origami and kirigami mechanics underpin innovative structural geometries with high reconfigurability and deployment efficiency. Origami techniques such as Yoshimura buckling patterns are used in pneumatic casings and deployable modules, enabling structures with large stroke-to-weight ratios. Tensegrity-inspired origami architectures further balance compliance and strength in space deployables. Kirigami skins—films with cut patterns yielding directional friction and elastic responses—enrich pneumatic actuators with locomotion and gripping functions. A kirigami-skinned pneumatic crawler, actuated pneumatically, was miniaturized to 45 g and contained power, actuation, sensing, and control in just 25 mm^3, demonstrating untethered soft robotic deployment capabilities [26].

Soft Continuum Manipulators: Soft continuum manipulators mimic invertebrate morphology, offering high degrees of freedom and seamless curvature. Typical implementations include octopus-inspired soft arms, elephant trunk-analog continuum actuators, and multi-module pneumatic manipulators (e.g., OctArm). These devices utilize pneumatic muscles or tendons within an elastic backbone to achieve fluid, continuous deformation [28]. Examples include OctArm (three McKibben modules), soft tentacle structures, and elephant trunk-like cable-actuated mechanisms. Advantages include safe interaction and compliance in cluttered environments, though challenges persist in achieving high positioning accuracy and advanced closed-loop 3D control.

Hybrid Rigid–Soft Body Designs: A hybrid architecture merges rigid support with soft compliant elements, yielding both structural strength and safe interaction capabilities. In articulated soft robotics, rigid "bones" support tendon-linked soft "muscles"—a vertebrate-inspired musculoskeletal model. The SIMBA project exemplifies this, combining modular continuum soft arms and reconfigurable grippers, achieving door-opening and object manipulation in robust scenarios [28]. Other hybrid designs include tendon-actuated robotic hands with variable stiffness skins, where tendons modulate fingertip rigidity for adaptive grasping. In medical continuum manipulators, rigid disk modules with soft springs and tendons yield compact, steerable tools appropriate for minimally invasive tasks [29].

This taxonomy of structural configurations illustrates how flexible manipulation mechanisms derive from both bio-inspired morphology and human-engineered innovation. Tendon-driven designs offer controllable deflection and stiffness; pneumatic and kirigami systems yield large deformations with simplified actuation; continuum robots provide rich dexterity; and hybrid systems balance compliance with structural integrity. Understanding these design archetypes is essential for selecting appropriate architectures in applications ranging from surgical robots to deployable field systems.

2.3 *Modeling and Simulation of Flexible Structures*

Precise modeling and simulation are pivotal in designing flexible manipulation structures, bridging theoretical principles and real-world implementation. This section explains four central modeling approaches: kinematic modeling (constant curvature and Cosserat rod theory), dynamic modeling, finite element modeling, and digital twins/virtual prototyping.

Kinematic Modeling Approaches. Constant curvature (CC) modeling approximates each continuum segment as an arc of fixed curvature, simplifying forward kinematics into closed-form solutions suitable for real-time control. The CC framework has been standard in continuum and soft robotics literature, enabling efficient mapping between actuator inputs and end-effector shapes [30]. In contrast, Cosserat rod theory treats the backbone as a continuous elastic rod subject to bending, shear, twisting, and axially loaded deformation. This more fundamental representation supports accurate modeling of spatial nonlinearities and external loads. It requires numerical integration of differential equations, typically solved via shooting or finite difference methods. Recent studies validate Cosserat-based models under multi-segment tendon actuation, achieving high correlation with experimental observations [31]. Hybrid approaches also employ discrete Cosserat formulations for enhanced computational tractability while retaining physical fidelity [32].

Dynamic Modeling Methods: Dynamic modeling captures time-dependent behavior, including inertial effects, damping, and external interactions. Approaches span lumped parameter models (mass-spring-damper chains), reduced-order models (e.g., CC dynamics), and full continuum dynamic models founded on Cosserat equations. A recent comprehensive review outlines the comparative merits and applications of these methods. Reduced-order CC-based dynamic models are often derived using Euler–Lagrange formalisms, applying simplifying assumptions (small deformations, low speed) to yield tractable but sufficiently accurate controllers. More detailed Cosserat-based dynamic simulations include complex nonlinearities and have begun to integrate machine learning elements to improve real-time viability [33]. Continual advancement aims to reconcile fidelity and real-time applicability.

Finite Element Modeling (FEM) for Soft Materials: The finite element method offers highly accurate representations of soft-body mechanics, solving continuum elasticity equations under large-deformation assumptions. Many soft robot designers employ FEM to simulate deformation under actuation or external loads, typically using tetrahedral or hexahedral meshes combined with nonlinear material models (e.g., neo-Hookean and Ogden). Recent work integrates differentiable FEM with simulation-based optimization, enabling gradient-based parameter tuning and direct sim-to-real pipelines for actuated soft structures [34]. These FEM frameworks are resource-intensive and often used during design and optimization stages, while simplified dynamic models are preferred during real-time deployment.

Digital Twins and Virtual Prototyping. Digital twins (DT) replicate physical structures within a virtual environment, enabling immersive prototyping, performance prediction, and real-time interaction. For soft robotics, DTs must model nonlinear, hysteretic, and time-variant behavior—including deformation, material aging, and sensor-actuator feedback loops. Case studies in soft grippers demonstrate successful DT implementations that capture hysteresis via Monte Carlo methods and permit reinforcement learning for control tuning. More broadly, DTs are employed in manufacturing applications, where sensor-driven virtual environments update in real-time to guide decision-making, reconfiguration, and predictive maintenance. Moreover, the union of digital twins with virtual reality or augmented reality interfaces affords immersive manipulation, teleoperation, and collaborative design, particularly valuable in medical or space-sensitive environments.

Each modeling tier delivers a unique balance of computational expedience and physical accuracy. CC-based kinematics and dynamics support fast control but at reduced fidelity; Cosserat rod models enhance physical representation across wider conditions; FEM offers precision during the design phase; and digital twins bridge design and operation with real-time adaptability.

Choosing among these methods depends on application demands, required control fidelity, computational resources, and real-time constraints. For example, a teleoperated surgical continuum robot may employ Cosserat dynamics for control fidelity, while a soft gripper for packaging utilizes FEM-based design and a DT for calibration and adaptation.

Continuing trends include: real-time hybrid modeling combining Cosserat and reduced-order models; machine learning enhancements to accelerate simulation; and immersive DT environments for end-stage deployment and reconfiguration.

2.4 *Integration Design*

2.4.1 Material–Structure Integration

Material selection underpins flexible manipulation structure performance. Common elastomers (e.g., silicone and polyurethane) exhibit large-strain capability (300–900%) and biocompatibility, suitable for safe contact and human–robot interaction [35]. Hydrogels and ionic polymers are chosen for applications requiring water-actuation or low-voltage responsiveness, useful in biomedical and wearable devices. Thermoplastic shape memory polymers (SMPs), by contrast, are ideal when stiffness change or self-folding structures are needed due to their glass transition-driven modulus shift.

Variable stiffness is critical to balance compliance with load-bearing capabilities. Techniques include layer jamming, granular jamming, and phase-change composites. For instance, vacuum-activated layer jamming enables safe-rigid state shifting. Granular jamming in soft grippers improves grip force while remaining gentle. Smart composites embedding low-melting-point alloys or SMPs transition stiffness with thermal stimuli. Electrorheological fluid–PDMS sandwiches offer >1500% stiffness change in <65 ms under electric field. Such rapid adaptability merges flexibility with structural resilience. Integration of sensors or actuators within soft structures enables multifunctionality. Textile platforms embed SMA wires for co-acting structure and sensing (e.g., soft gripper that senses bend angle). Dielectric elastomer actuators (DEAs) employ compliant electrodes on elastomer films to provide actuation and sensory feedback, exhibiting high energy density (~3 J/g), fast response (<1 ms), and large strain (>1000%) [35]. Magnetic-particle-laced SMP composites allow Joule-heated stiffness tuning with programmable actuation segments. Such composites realize structural, sensory, and actuation integration in a compact form factor.

2.4.2 Actuation Strategies for Structural Design

Pneumatic/Hydraulic Actuation and Structure Interaction: Pneumatic artificial muscles, like McKibben actuators, generate radial expansion under pressure and high force-to-weight ratios. Their compliance arises from fluid-structure interactions in elastomer chambers. Hydraulic analogs offer higher force but remain less common due to leakage issues. Hybrid designs combine pneumatics with stiffness-modulation methods (e.g., granular jamming) to decouple shape and rigidity control.

Shape Memory Alloys (SMA) and Dielectric Elastomers (DEA): SMAs use thermally induced phase transitions to produce contraction; Nitinol wires achieve high force densities and are used in tactile robots and assistive grippers [36]. SMA–textile composites can actuate self-folding structures and detect deformation. DEAs—thin elastomer films with compliant electrodes—actuate via electrostatic pressure, yielding fast (<1 ms), large-strain actuation with energy densities comparable to muscle (~3 J/g). They are widely used in soft grippers, anthropomorphic robots, and haptic devices.

Magnetic and Thermal Responsive Materials Integrated into Design: Magnetic-responsive composites—polymers imbued with magnetic particles—shift shape in external fields. Multi-segment composites wire-heated for stiffness and magnetized permit segmental control and high payload manipulation. SMPs and low-melting alloys embed thermal-response: SMP-grippers self-fold and lock under heat; composite SMP skins enact shape during deployment. Liquid crystal elastomers allow reversible thermal deformation and caterpillar-like motion [37].

Distributed vs. Centralized Actuation in Soft Manipulators: Actuation architecture significantly influences system performance. Centralized actuation concentrates pumps or motors externally, reducing onboard mass but increasing tethering and latency. Distributed actuation, with embedded or local triggers (DEAs, SMAs, and jamming layers), achieves autonomy and rapid response. Textile platforms and modular DEAs demonstrate distributed actuation enabling local shape/stiffness control. Magnetic composite manipulators with segmental coils exemplify spatial actuation complexity. Trade-offs involve power delivery, heat dissipation, and system complexity.

Material–structure integration and actuation co-design form the foundation for achieving intelligent, compliant, and adaptive robotic manipulators. Designers must select materials aligned with goals—whether compliance, stiffness-tuning, rapid actuation, or integrated sensing. Actuators must complement structural features: pneumatics offer fluid compliance; SMAs and DEAs balance compact actuation with force and speed; magnetic and thermal materials enable programmable structure. Embedding distributed actuation supports autonomy and responsiveness. Future research will likely explore multifunctional composites (e.g., magneto-active SMP), self-healing materials, and integrated energy harvesting, leading to untethered, intelligent soft manipulators.

2.5 Design Methodologies and Optimization

Flexible manipulation structures demand sophisticated design workflows that reconcile multiple performance objectives—be they structural, energetic, functional, or contextual. This section reviews four complementary methodologies: multi-objective design optimization; topology optimization specific to soft structures; machine learning-enhanced structure generation; and holistic co-design of structure and control mechanisms.

Multi-Objective Design Optimization. Multi-objective optimization (MOO) techniques are essential in soft robotics, where conflicting requirements such as stiffness, workspace, and energy consumption must be simultaneously balanced. Typical MOO formulations seek Pareto-optimal solutions under complex, often non-convex design spaces [38]. A representative example comes from parallel kinematic machines, where genetic algorithms optimize link dimensions and joint configurations to maximize workspace while enhancing stiffness. In exoskeleton design, studies optimize actuator placement using stiffness ellipsoid metrics within ergonomic workspace constraints [39]. For energy efficiency, dual-arm industrial manipulators employ metaheuristics (e.g., GA, PSO) to balance time, energy consumption, and torque requirements.

These frameworks typically utilize weighted sum or Pareto front strategies to explore trade-offs, using objective functions such as workspace coverage, mechanical stiffness, torque summation, or energy metrics. Decision-makers then select optimal solutions aligned with application priorities, considering manufacturing and sensing constraints.

Topology Optimization for Soft Body Design. Topology optimization allocates material within a design domain to maximize performance criteria under specified loads and constraints. While traditionally used in rigid components (aerospace and automotive), its adoption in soft robotic design is growing due to additive manufacturing (AM) enabling intricate geometries. Density-based methods (e.g., Solid Isotropic Material with Penalization—SIMP) optimize compliance and stiffness-to-weight ratios, producing organic shapes suited for

3D printing [40]. These techniques have been used to derive soft continuum backbones with spatial stiffness profiles for compliant robotic arms. Yet, manufacturing constraints—e.g., minimal feature size—must often be integrated in the optimization loop. Hybrid approaches now embed topology within generative design systems, iterating between AM capabilities and structure-function metrics.

Machine Learning-Assisted Structure Design: Machine learning (ML), particularly generative models and reinforcement learning, now accelerates soft structure design by automating shape discovery under performance constraints. Frameworks like "MLGen" combine ML with topology optimization to produce lightweight, high-performance structural candidates for AM. Benchmark platforms like SoftZoo enable co-optimization of soft robot morphology and control policy across ecosystems—e.g., various terrains and objectives—via differentiable design representations [41]. These enable automatic discovery of body–brain combinations that exploit compliance while achieving task-level goals. Generative design tools embedded in CAD suites (e.g., Autodesk Inventor) integrate gradient-based and genetic solvers, producing manufacturable structures adhering to AM guidelines, while satisfying stiffness and mass objectives.

Robotic Co-Design: Structure + Controller Co-Optimization: The future of soft robotics lies in co-design—simultaneous optimization of both physical morphology and control policies. Co-design addresses the body-brain coupling present in soft systems, where mechanical properties dramatically influence control effectiveness. Recent work by Sato et al. develops a differentiable topology optimization integrated with neural network controllers to optimize locomoting soft robots under terrain uncertainties. Mertan and Cheney advocate the use of modular controllers to decouple morphology-controller dependencies, improving co-optimization robustness [42].

These advances mark a departure from traditional design sequences by integrating physical and control layers during conceptualization, offering robots that are structurally and behaviorally synergistic.

2.6 *Case Studies and Engineering Applications*

This section explores how flexible manipulation structures and advanced design methodologies are applied across critical domains: biomedical, industrial, aerospace, and marine operations. Each domain highlights unique functional demands and design outcomes.

Biomedical: Endoscopy and Rehabilitation Exoskeletons: Flexible manipulators are revolutionizing minimally invasive surgery (MIS). A notable case is a hybrid-driven soft surgical robot optimized using FEA and response surface optimization to minimize radial expansion and actuation energy for dexterous RAMIS tasks [43]. The study leveraged silicone properties experimentally validated before optimization, yielding a prototype with enhanced precision and lower actuation forces, further tested against anatomical phantoms.

In rehabilitation, shoulder exoskeletons using compliant-beam parallel architectures have undergone actuator placement optimization via stiffness ellipsoid volumes to maximize compliance during human interaction. Here, workspace, stiffness distribution, and ergonomic constraints guided design, validating optimized configurations experimentally.

Industrial: Automated Soft Grippers for Irregular Objects: In manufacturing, soft grippers are valued for handling irregular items like food, textiles, or complex geometries. For instance, granular jamming soft grippers have been optimized using multi-objective GA methods to reconcile grip strength with gentleness—a balancing act of stiffness and compliance.

Additionally, generative design tools combined with optimization algorithms have produced lightweight, tunable gripper structures for robotic arms, achieving 27% mass reduction and 95% energy gain in specific load-carrying frames [44]. Integration with topology optimization tailored components to manufacturing constraints, yielding robust yet compliant end-effectors.

Aerospace: Deployable Morphing Structures: Aerospace leverages flexible architecture for lightweight, deployable structures. Topology optimization has been widely used in aerospace AM parts to reduce mass while maintaining stiffness. Although soft body actuators see limited use in primary systems, hybrid rigid–soft morphing skins remain key for adaptive aerodynamics and damage tolerance. Preliminary co-design approaches incorporate soft skin-material and deployment controllers to enable shape change during flight, taking inspiration from morphing wing concepts in drones and satellites.

Marine: Deep-Sea Delicate Sampling with Soft Arms: In marine biology and underwater inspection, soft manipulators offer unparalleled delicacy and deformability. Continuum soft arms—modeled on octopus limbs and actuated via pneumatics—are deployed for fragile specimen sampling in deep-sea conditions. These systems integrate pressure-tolerant materials, embedded sensors, and tendon-driven segmentation to ensure safe interaction with biological specimens without damage to marine life or equipment. Co-design algorithms ensure that arm geometry and control are optimized for depth, fluid dynamics, and payload recovery—critical for near-real-time biological studies.

This dual-section synthesis demonstrates that designing flexible manipulation systems today is a multi-faceted process—spanning structured optimization, generative methods, and holistic co-design approaches. Across domains, these methodologies have led to successfully deployed systems: from internal surgical manipulators to industrial grippers and environmental exploration arms. Future advancements are expected in real-time co-design loops, integrated physical-digital twins, multifunctional composites, and self-adaptive systems responsive to evolving operational contexts.

For research teams and engineering practitioners, selecting methodologies depends on application-specific priorities: medical devices often emphasize precision and biocompatibility; industrial systems focus on throughput and energy efficiency; aerospace requires deployability and lightweight resilience; marine platforms stress robustness under environmental uncertainty. The cross-pollination of soft robotic design techniques will continue to yield innovations across sectors.

2.7 Challenges and Future Directions

Flexible manipulation structures exhibit immense promise, yet they face significant technical hurdles that must be addressed via interdisciplinary innovation.

Trade-offs Between Flexibility and Control Precision: High compliance often complicates precise positioning and task execution. Soft materials' low stiffness makes accurate modeling and control challenging, requiring sophisticated sensing and adaptive algorithms. The absence of standardized soft–rigid interfaces exacerbates this issue, hindering predictable energy and information transfer across modules. To reconcile dexterity and compliance, researchers are developing custom control architectures and embedded proprioceptive sensor networks tailored specifically for flexible mechanisms. Additionally, rigid–soft hybrids using discrete stiff elements offer a compromise, improving controllability without sacrificing adaptability.

Durability and Fatigue Resistance of Soft Structures: Under repeated deformation, soft materials experience fatigue, wear, and permanent deformation, threatening long-term reliability. Studies on architectured elastomeric systems reveal accumulation of plastic deformation under cyclic loading, impairing precision over time. Material degradation—including micro-tears and crack propagation—is aggravated in extreme environments, requiring experimental quantification of fatigue limits and nonlinear wear behaviors. To mitigate these issues, research focuses on self-healing composites with embedded stimuli-responsive agents (e.g., microcapsules) that autonomously repair minor damage, enhancing resilience. Additionally, tensegrity and hybrid structural designs distribute stress and improve impact tolerance, promoting system-level durability.

Scaling Challenges in Micro and Macro Domains: Dimension presents unique challenges in flexible manipulators. Microscale systems face fabrication precision limitations, surface-dominated forces, and unreliable fluidic actuation. Conversely, macro-scale designs must overcome material toughness, actuation pressure, and weight constraints. Fabrication at microscales is advancing via multi-material 3D printing, but resolution trade-offs remain. At large scales, achieving uniform material properties and consistent stiffness is difficult, and sustaining soft bodies under their own weight necessitates variable stiffness and hybrid rigid–soft architectures.

Toward Autonomous Self-Reconfigurable Structures: Autonomous self-reconfiguration embodies next-generation flexibility—structures that can adapt functionally and physically without external guidance. Modular robotics has made progress, such as "Roombots" and lattice-based Cube-type modules, yet scaling to hundreds or thousands of units remains tough due to interface strength, distributed control, and inter-module communication challenges. Biological inspirations like cellular repair, amoeboid reassembly, and developmental morphogenesis suggest future directions. Progress in resilient and self-healing soft robots indicates pathways where structures autonomously detect damage, heal, and reconfigure functional connectivity. Effective advancement requires integrated hardware–software ecosystems supporting local autonomy, environmental sensing, and adaptive control—suggesting a future of self-aware, self-building robotic matter.

2.8 Summary

This chapter section examined both obstacles and innovations in the development of flexible manipulation structures. Key challenges—ranging from control precision, fatigue resilience, and size scaling, to autonomous self-reconfiguration—highlight the complex interplay between materials, mechanics, and intelligent control systems.

To address control limitations, research must focus on dedicated control frameworks and refined sensing–actuation integration. Durability challenges necessitate fatigue quantification, self-healing materials, and stress-distributing designs. Micro- and macro-scale implementations require advances in fabrication and hybrid stiff-soft mechanisms. Achieving autonomous self-reconfiguration will involve breakthroughs in modular interfaces, distributed intelligence, and material-level responsiveness.

Looking forward, the pathway to truly flexible, resilient, and intelligent manipulation systems lies in converging adaptive materials, multifunctional composites, embedded sensing networks, decentralized control architecture, and bio-inspired meta-materials. Such synergy will enable robotic systems that are not only functionally flexible but also structurally self-adaptive and context-aware, aligning robotic capabilities more closely with the versatility observed in natural systems.

3 Materials for Flexible Manipulators

Actuators are fundamental components in many industrial applications such as robotics, automobiles, military facilities, and consumer products, traditionally utilizing pneumatic, hydraulic, and electromagnetic technologies. These conventional actuators are known for their energy efficiency but are limited by low work and power densities due to their reliance on hard and heavy materials. Modern advancements aim to create actuators that are smaller, lighter, faster, and softer, in line with the ethos of reducing energy consumption, increasing speed, and improving safety and flexibility through the use of compliant materials similar to those found in living organisms.

The next generation of actuators for soft robots requires a significant departure from traditional technologies, building on extensive research achievements. Introduced in the late 1990s, smart materials have revolutionized actuators with their ability to alter shape, stiffness, position, and other mechanical characteristics in response to environmental stimuli like temperature and electromagnetic fields [45]. These materials belong to solid state actuator technology, which is characterized by high energy and power densities, showcasing materials such as shape memory alloys, piezoelectric, magnetostrictive, electrostrictive, and electroactive polymers (EAP).

EAPs have garnered attention due to their capability for large deformations, making them highly suitable for a myriad of applications that require mechanical properties similar to biological muscles. The soft nature of these materials in EAPs allows for a high degree of freedom in movement without the need for joints or hinges, lending themselves well to autonomous soft robots that are composed of flexible materials without rigid components. These robots can absorb energy during impacts, interact safely with humans, and navigate through complex terrains, highlighting the potential to enhance the capabilities of traditional robotic systems significantly [46].

Recent advancements have also been facilitated by innovations in additive manufacturing, such as 3D printing, which supports the free-form fabrication of complex structures. This has opened up new possibilities for soft robots capable of sophisticated actions, such as jumping and complex 3D movements [47]. Current reviews have highlighted the importance of various performance criteria, including compliance, scalability, energy efficiency, and operational range, which are essential for selecting suitable soft actuator materials for specific applications.

The ongoing development of soft materials, inspired by biological and natural systems, is pushing the boundaries of what is possible with actuator technologies, promising more adaptive, efficient, and versatile solutions for modern industrial applications.

Gels are colloidal suspensions or solutions that undergo a transformation into a solid state while retaining a significant amount of solvent. They are defined as elastic solids composed of at least two components: a polymer that forms a three-dimensional network either through physical or chemical bonds and a liquid medium that ensures the gel's elasticity [48]. This network not only retains the liquid but also imparts solidity to the gel, allowing it to sustain large deformations in response to environmental changes, thereby exhibiting certain mechanical properties, such as yield strength and elasticity.

Physical and chemical gels differ in their cross-linking networks. Physical gels are reversible and can transition between solid-like and liquid-like states under specific conditions, making them suitable for applications that benefit from temporary structures, like drug delivery systems. Chemical gels, created through irreversible covalent bonds, provide more permanent and stable structures used in various industrial applications.

Hydrogels, a special type of gel, consist of polymeric chains that are swollen in water and form three-dimensional network structures [49]. These networks are not only insoluble but also highly hydrophilic, making hydrogels ideal for biomedical applications, such as tissue engineering, drug delivery, and implants. Their hydrophilic and biocompatible nature allows for extensive use in both laboratory and clinical settings.

Another fascinating gel type is the organogel, which involves an organic liquid phase stabilized by a cross-linked polymer network. Despite their largely liquid composition, organogels exhibit solid-like rheological behavior, making them attractive for various forms of drug delivery.

Ionogels are a unique class of gels that incorporate ionic liquids within a polymer matrix. These gels combine the properties of ionic liquids—such as high ionic conductivity and thermal stability—with the mechanical properties of the polymer network, leading to applications in electrochemical devices and smart material systems.

Stimuli-responsive gels, which can undergo transitions from gel to solution or gel to solid upon application of external stimuli, represent a versatile class of materials [50]. These gels can respond to a variety of stimuli, including temperature, pH, and electromagnetic fields, making them useful in applications ranging from environmental sensing to smart actuators in biomedical engineering.

3.1 pH Responsive

pH-responsive hydrogels are a specialized category of stimuli-responsive materials that exhibit significant swelling or deswelling based on pH changes in their environment [49]. These hydrogels are particularly valuable in biomedical fields where body environments often experience pH fluctuations due to normal physiological processes or pathological conditions. The unique characteristic of these hydrogels comes from ionizable groups within the polymer structure that respond dynamically to pH variations, altering the hydrogel's volume accordingly.

Two main types of pH-responsive hydrogels exist, differentiated by the nature of their ionizable groups: anionic and cationic. Anionic hydrogels swell when the pH exceeds their p*K*a due to the repulsion between negatively charged groups, while cationic hydrogels swell at pH levels below their pKa because of the protonation of their groups. This behavior is crucial for applications like drug delivery systems, where the release of therapeutic agents can be precisely controlled by the pH of the surrounding tissue.

Furthermore, the hydrogels' porous nature can be finely tuned to modulate flow and permeability in various applications, from membrane technology to tissue scaffolds. This is achieved by manipulating the cross-linking density and the nature of the polymer components, which react to pH changes by opening or closing the network pores.

Natural polymers, such as chitosan, alginate, and cellulose derivatives, also show pH-responsive behaviors due to their inherent properties of hydrophobic interactions or charge-based linkages [51]. These natural hydrogels are not only effective in swelling and deswelling in response to pH but are also advantageous for environmental and biomedical applications due to their biocompatibility and biodegradability.

Innovative designs of pH-responsive hydrogels utilize cellulose nanocrystals and other nanomaterials to enhance mechanical and dielectric properties, making these hydrogels more effective and versatile in smart device applications, including soft actuators. The dynamic response of these hydrogels to pH changes is essential for developing responsive systems that can operate in complex biological or environmental conditions.

3.2 Temperature Responsive

Temperature-responsive hydrogels are categorized into two main types based on their response to thermal changes: those that swell at temperatures below a lower critical solution temperature (LCST) and those that form gels above an upper critical solution temperature (UCST). These materials are pivotal in developing smart devices due to their ability to undergo significant physical changes in response to temperature variations, making them ideal for applications like drug delivery, tissue engineering, and smart textiles.

LCST-type hydrogels typically exhibit a shrinkage or deswelling at temperatures above their LCST, which can be leveraged to release encapsulated substances such as drugs in a controlled manner. For instance, hydrogels made from poly(*N*-isopropylacrylamide) (PNIPAAm) are well-known for their rapid response to temperature changes around the human body temperature, making them suitable for biomedical applications.

Conversely, UCST-type hydrogels expand and form stronger gel structures when heated above their UCST. These materials are less common but have potential in applications where heat-induced deployment of materials is necessary, such as in injectable hydrogels that solidify at body temperature for tissue scaffolding or wound healing.

To enhance the mechanical properties and responsiveness of these hydrogels, modifications with various co-monomers and cross-linking agents are employed. These modifications help tailor the physical properties of the hydrogels, such as stiffness and porosity, which are critical in specific applications like microfluidic devices and responsive membranes.

Moreover, the integration of nanomaterials such as cellulose nanocrystals into hydrogels has led to the development of advanced composites with finely tunable properties. These composites can undergo significant changes in their mechanical strength in response to temperature, proving useful in structurally demanding applications.

Temperature-responsive hydrogels continue to inspire new technological advancements due to their dynamic nature and compatibility with various manufacturing processes, including photopolymerization and 3D printing. This adaptability allows for the creation of complex structures tailored to specific functional needs, from single-cell handling systems to large-scale industrial processes.

3.3 *Photoresponsive*

Photoresponsive polymers react to light through various bonding and degradation reactions, adhering to the Planck-Einstein relation, which correlates the reaction rates closely with light energy and irradiance. Such properties make them suitable for manipulating hydrogel structures, although their use in vivo is limited due to the poor tissue penetration of UV and visible light. These limitations restrict their applicability mainly to ex vivo systems or superficial treatments.

Advancements in near-infrared sensitive reactions have expanded their utility, allowing deeper tissue interactions. One innovative application involves a supramolecular actuator utilizing host-guest interactions, where a hydrogel embedded with α-cyclodextrin and azobenzene derivatives can reversibly alter its shape under specific light wavelengths. Such dynamic behavior is crucial for targeted therapeutic applications where precise control over drug release is necessary.

Furthermore, the integration of materials like reduced graphene oxide into hydrogels (e.g., PNIPAAm/rGO) enhances their photothermal responsiveness. These hydrogels can significantly contract under light exposure, offering practical uses in light-driven actuators that mimic natural movements, such as the bending and twisting observed in plant tendrils. This biomimetic approach is not only relevant for developing soft robotic applications but also for environmental sensing and interactive devices.

In terms of hydrogel responsiveness, ongoing research focuses on improving the kinetics of these materials to enhance their practicality in real-world applications. The inclusion of humidity-responsive elements, for instance, demonstrates the potential for creating actuators that can operate autonomously in varying atmospheric conditions without direct contact, crucial for applications requiring non-invasive operation.

3.4 *Electrical Responsive*

Electrical-responsive gels have been a significant area of research, initiated by pioneering studies using PAMPS hydrogel immersed in a surfactant solution [52]. This gel demonstrated mobility by walking through a solution with the application and reversal of a 20 V potential at 2-s intervals. This mobility results from a reversible complexing of surfactant molecules in the gel under an electric field, leading to contraction and expansion.

Advancements in this field have also included the use of nonionic polymer gels and dielectric liquids to form systems like poly(vinyl alcohol)/dimethylsulfoxide (PVA/DMSO) [53] and poly(vinyl chloride)/di-*n*-butylphthalate (PVC/DBP) [54], which can respond electrically with strains up to 8% and bend up to 180° in milliseconds. Further developments in PVA/DMSO gels have achieved controlled motion in air, suggesting their potential for high-speed applications and dynamic environments.

One innovative application of electric-responsive gels is in the development of soft hydrogel walkers for cargo transport. These hydrogel systems, which mimic biological motion, offer new avenues for creating soft robotic systems with simple designs capable of complex tasks. Additionally, ionogel electrolytes have been developed for use in electrochemical actuators, providing low-voltage control, high electrochemical stability, and silent operation.

These gels have also found niche applications in reconfigurable lenses for optics, where electrical stimuli adjust the lens without any moving parts, maintaining optical transparency and functionality. This adaptability illustrates the broad potential of electric-responsive gels beyond traditional applications, expanding into fields requiring precise and variable responses to electrical stimuli.

3.5 Chemical Responsive

Chemically responsive hydrogels are engineered to swell or degrade in response to specific target molecules, allowing for a broad range of applications, including drug delivery and sensor development. These hydrogels utilize various mechanisms to detect and respond to chemicals. For instance, glucose-responsive hydrogels can significantly impact diabetes treatment by responding to blood glucose levels, thereby releasing insulin as needed without the need for frequent injections.

Glucose-responsive systems typically utilize mechanisms such as the enzymatic breakdown by glucose oxidase or the binding capabilities of lectins like concanavalin A, which selectively binds glucose, triggering a swelling response. Another approach involves phenylboronic acid (PBA), which forms reversible bonds with diols in glucose, altering the hydrogel's charge and hydrophilicity to induce swelling.

Beyond glucose, hydrogels can be tailored to respond to other biomolecules through molecular imprinting. Molecularly imprinted polymers (MIPs) are designed with specific cavities that match target molecules, functioning similarly to antibodies or aptamers. These are promising for applications like cost-effective sensors and targeted drug delivery systems.

Moreover, advancements in materials such as graphdiyne-based electrochemical actuators show potential in converting electrical to mechanical energy efficiently, with applications in fields like artificial intelligence. These materials offer high electromechanical transduction efficiency and are capable of operating across a wide range of frequencies, highlighting their versatility and robustness in various technological applications.

3.5.1 IPMCs

Electroactive polymers (EAPs) are classified into two main types based on their activation mechanisms: electronic and ionic. Electronic EAPs, such as electrostrictive, electrostatic, piezoelectric, and ferroelectric polymers, typically require high activation fields that approach the material's breakdown level. In contrast, ionic EAPs, which include gels, polymer-metal composites, conductive polymers, and carbon nanotubes, operate at much lower voltages. However, maintaining moisture within these materials is crucial, and it is challenging to sustain displacements induced by direct current, especially in materials other than conductive polymers.

Ionic Polymer-Metal Composites (IPMCs) represent a significant category of ionic EAPs. An IPMC consists of a core ion-conductive polymer membrane enriched with mobile ionic salts and plated with noble metal electrodes. The bending motion in IPMCs is driven by the redistribution of ions toward the oppositely charged electrode when an electric field is applied. Discovered in the early 1990s, IPMCs have seen various enhancements in performance through the development of metal electrode deposition techniques and the selection of ionic polymers that include specific counter cation species, which improve actuation qualities.

The actuation mechanism of IPMCs relies heavily on the ion transport and electromechanical properties of the ionic polymers, necessitating the design and fabrication of high-performance actuators with specialized micro-/nano-structured polymer matrices. These structures often include hydrophilic nano-channels and microphase-separated morphologies to enhance performance.

Despite their potential, typical ionic polymer actuators face challenges such as stress relaxation under sustained inputs, low force output, durability issues, environmental concerns, and high production costs associated with the use of fluorinated polymers and noble metals. To address these drawbacks, recent research has focused on developing alternative materials like sulfonated hydrocarbon polymers, block copolymers, functionalized cellulose-based biopolymers, and carbon-based nanocomposite membranes.

IPMCs have been utilized in various applications, from actuators and sensors in flexible braille displays and soft bio-inspired fins to biomimetic devices like synthetic Venus flytraps and flow measurement sensors. Moreover, advancements in IPMC technology have led to the development of actuators capable of complex, multi-shape memory effects that respond to multiple external stimuli, such as electrical and thermal inputs, enabling sophisticated movements like simultaneous bending and twisting.

3.5.2 Conducting Polymers

Conductive polymers (CPs) are distinguished by their reversible counter-ion insertion and expulsion during redox cycling, marking them as either semiconductors or insulators. Their conductivity significantly increases upon oxidation, achieved either through exposure to oxidizing agents or electrochemical methods [55]. During oxidation, CPs swell due to the insertion of bulky ions, making them flexible yet mechanically robust, ideal for applications in electrically controllable soft actuators.

CP actuators are developed in various forms, such as films, filaments, yarns, and textiles, and are operable both in liquids and air. Extensive research on CPs, particularly on polypyrrole (PPy), polyaniline, polythiophene, and poly(3,4-ethylenedioxythiophene), has established them as stable due to their aromatic structures. PPy, for instance, is favored for actuator applications because of its ease of electrodeposition on metal electrodes and its ability to form highly conductive and durable films.

The actuation mechanism of CPs involves the transport and swelling caused by ion insertion, greatly influenced by the type of ion used. Smaller ions allow for more significant and reversible volume changes, enhancing the CP's actuation capabilities. This reversible expansion and contraction make CPs suitable for various actuator configurations, leading to applications in soft robotics, medical devices, and more. These actuators can mimic natural movements, offering potential in bio-inspired applications.

Despite their advantages, CPs face challenges like stress relaxation under continuous operation, environmental sensitivity, and high production costs due to the complex manufacturing processes involved. Innovations continue to focus on overcoming these drawbacks through material enhancements and novel fabrication techniques.

3.5.3 CNTs and Graphene

Carbon-based materials like carbon nanotubes (CNTs) and graphene are versatile due to carbon's ability to form various bonds. Single-wall carbon nanotubes (SWNTs), for example, exhibit significant dimensional changes upon charging and discharging, making them excellent for converting electrical to mechanical energy in soft actuators [56]. These materials offer high surface area, exceptional mechanical properties, and a non-faradic electrochemical response, contributing to their durability and efficiency.

The actuation in SWNTs relies on quantum-chemical effects triggered by the double-layer charge at the interface between the nanotube and the solution. By altering the applied potential, electronic charges accumulate in the SWNT, balanced by ions at the carbon nanotube/solution interface, creating a double layer that induces dimensional and stress changes in the material.

For multi-wall carbon nanotubes (MWNTs), the mechanism differs slightly. Their actuation involves expansion during charging due to repulsion between electrical double layers associated with each nanotube. This expansion can lead to higher actuator responses than traditional materials.

Challenges remain, particularly in optimizing these materials' assembly to fully harness their mechanical properties and improving their response times. Innovations in spinning MWNTs have led to unique mechanical actuations involving coupled rotation and axial contraction, driven by electrolyte ion influx/release during electrochemical processes or thermal expansion/contraction of a guest material.

Despite the potential for high stress generation and giant work capacities, significant hurdles such as maintaining wetness and managing slow response rates need addressing to maximize practical applications. New developments like all-solid-state MWNT yarns and plied yarn muscles that can operate in ambient air are promising advances that mitigate some of these challenges, broadening the potential applications of these advanced materials in soft actuators and beyond.

3.5.4 Dielectric Elastomers

Dielectric elastomers (DEs) function as soft and flexible electrical insulators that can exhibit mechanical motions in response to electrical stimuli. When a dielectric membrane is subjected to an electric voltage across its thickness, it experiences electrostatic forces—commonly known as Maxwell stress—which causes it to compress in thickness and expand in area [57]. This property has been utilized over the past three decades to develop DE-based electroactive polymers (EAPs) that are rubbery with compliant electrodes, displaying a significant electromechanical response to applied electric fields.

The strain induced in DEs is proportional to the square of the electric field, enhanced by the material's dielectric constant, and inversely proportional to its elastic modulus. Polymers with high dielectric constants subjected to high electric fields can generate substantial forces and strains, making them highly effective for actuation applications. DEs are incorporated in various devices, including linear artificial muscles, insect-inspired robots, and flapping-wing robots, owing to their ability to convert electrical input into precise, controlled movements.

However, DEs require high voltages for activation, which presents challenges, such as voltage breakdown, the need for efficient shielding, packaging concerns, and the complexity of miniaturization and driver configuration in device implementations. Innovations aim to enhance the dielectric constant while reducing the elastic modulus to lower the required actuation voltages. Recent developments include all-polymer percolative composites that exhibit very high dielectric constants.

Furthermore, DEs are also being explored for their potential in sensors and energy harvesters. Their ability to undergo substantial deformations has led to innovations like contractile polymer-based electromechanical linear actuators that exploit DEs in novel configurations for axial contractions and radial expansions.

Despite these advances, significant challenges remain in fully utilizing the mechanical properties of DE materials to optimize performance, particularly in achieving high actuation amplitudes without compromising structural stability. All-polymeric actuators with solid composite elastomer electrodes are being studied to address these limitations effectively.

Emerging applications of DEs include their integration into next-generation flexible displays, touchscreen panels, and smart optics, where their optical transmittance and high dielectric constant are particularly valuable. Transparent DEs with over 90% transmittance have been reported, supporting their use in visually clear applications where flexibility and high performance are crucial.

3.5.5 Shape Memory Polymers

Shape memory polymers (SMPs) are remarkable smart materials that can transform from a temporary deformed state back to their original configuration in response to specific external stimuli, such as heat, light, electric and magnetic fields, or moisture [58]. These materials consist of two primary components: switching units that control shape fixity and recovery, and netpoints that define their permanent shape. The netpoints can be either covalently bonded polymer segments within cross-linked networks or physical cross-links seen in block copolymer-based SMPs.

Two-way SMPs (2W-SMPs), a special category, are particularly interesting because they expand when cooled and contract when heated, a behavior opposite to common physical responses. This reversible action allows them to cycle through changes continuously once activated. Initial applications of 2W-SMPs included liquid crystalline elastomers and have since expanded to include various polymers like poly(cyclooctene) and polyurethane, among others. These materials have been utilized in creating soft actuators, morphing structures, and even biomedical devices like artificial muscles due to their dynamic properties.

However, traditional 2W-SMPs have faced challenges such as limited operational temperature ranges and lower tensile strength and stiffness. Recent advancements have focused on overcoming these limitations by integrating inorganic nanofillers to enhance their mechanical properties and functional versatility. For example, new composite networks that combine polycaprolactone with cellulose nanocrystals have shown improved mechanical properties and responsiveness to thermal and moisture cues.

Innovative developments have also led to SMPs with triple-shape or multiple-shape memory capabilities that exceed the conventional dual-shape functionalities, broadening their application spectrum. Furthermore, the integration of plasticity into SMP networks enables permanent and nonrecoverable shape changes in response to external forces, facilitating the creation of complex structures and multifunctional devices. This dual capacity for temporary elasticity and permanent plasticity opens new avenues for the use of SMPs in areas ranging from robotics to customizable medical implants.

3.5.6 Biopolymers

Biopolymers, derived from renewable sources, are pivotal for harmonizing future technologies with our environment. Utilizing biopolymers like cellulose, chitin, bamboo, and others is crucial to ensuring sustainable development. These materials have proven indispensable in various sectors due to their lightweight, biocompatibility, and robust mechanical properties.

Recent advancements have seen cellulose emerge as a key player in the realm of electroactive polymers, notably in the development of cellulose-based electroactive paper (EAPap). This innovative material showcases dual functionalities—piezoelectric and ionic. While Piezopaper, the piezoelectric form of EAPap, demonstrates high-frequency deformations, its ionic counterpart excels in producing significant deformations at lower frequencies through ion migration within the cellulose matrix. This is enhanced by integrating ionic liquids into the cellulose, boosting its ionic transport capabilities without the need for encapsulation, making it suitable for applications in biomimetic actuators.

The structural composition of cellulose, from its crystalline to amorphous forms, lends itself to high stability and strength, thanks to the robust hydrogen bonding in its nanocrystalline cellulose (CNC) and cellulose nanofiber (CNF) forms [59]. Known as nanocellulose (NC), this material is not only sustainable but also features excellent physical and chemical attributes like high elastic modulus and transparency, making it highly versatile across various applications—from automotive and packaging to advanced electronics and medical devices. The expansive use of NC underscores its potential as a "post-carbon" material, pivotal for future technological and environmental advancements.

This section comprehensively explores the array of materials used in soft actuators, such as gels/hydrogels, ionic polymer-metal composites (IPMCs), conducting polymers, carbon nanotubes/graphene, dielectric elastomers, shape memory polymers, and biopolymers. It delves into their preparation, synthesis, performance characteristics, and the challenges they face, including how their softness is quantified by comparing their elastic moduli to those of more conventional actuators, typically within the range of gigapascals.

Significant attention is paid to the unique properties of stimuli-responsive gels/hydrogels that enable large deformations under varying conditions—pH, temperature, light, chemical, electric, and magnetic fields. Despite their advantages, issues such as maintaining wetness, slow response times, and hysteresis are noted as challenges that need addressing. Dielectric gels are highlighted for their high dielectric constants, allowing significant electrostatic forces to be generated under electric fields.

This part also reviews advances in IPMCs and conducting polymers enhanced by ionic liquids, as well as yarned carbon nanotubes that offer promising contraction capabilities. Dielectric elastomers are recognized for producing substantial strains under high electric fields, but there is a need for materials with higher dielectric constants to reduce required voltages and for the development of stretchable, flexible electrodes.

Further, advancements in shape memory polymers are discussed, particularly those with multiple-shape and reversible capabilities, ideal for creating tunable kirigami structures. Lastly, the role of biopolymers is emphasized; materials like nanocellulose are praised for their environmental friendliness, renewability, and biocompatibility, positioning them as pivotal elements in the next generation of "post-carbon" materials.

A deeper understanding of these materials, coupled with innovative design strategies, could significantly broaden their application in fields, such as soft mechatronics, soft robotics, medical devices, haptic interfaces, flexible electronics, and smart optics.

References

1. Xie, J. (2019). *Dynamic modeling and control of flexible manipulators: A review*. School of Mechanical Engineering, Tianjin University.
2. Zhang, J. (2017). Review of soft-bodied manipulator. *Journal of Mechanical Engineering, 53*(13), 19. https://doi.org/10.3901/jme.2017.13.019
3. Shintake, J., Cacucciolo, V., Floreano, D., et al. (2018). Soft robotic grippers. *Advanced Materials, 30*(29), 1707035.
4. Ilievski, F., Mazzeo, A. D., Shepherd, R. F., et al. (2011). Soft robotics for chemists. *Angewandte Chemie International Edition, 50*(8), 1890–1895.
5. Yu, S. H., Chang, J. S., & Tsai, C. H. D. (2021). Grasp to see—Object classification using flexion glove with support vector machine. *Sensors, 21*(4), 1461.
6. Polygerinos, P., Wang, Z., Galloway, K. C., Wood, R. J., & Walsh, C. J. (2015). Soft robotic glove for combined assistance and at-home rehabilitation. *Robotics and Autonomous Systems, 73*, 135–143.
7. Wei, S. (2017). Design of a soft pneumatic robotic gripper based on fiber-reinforced actuator. *Journal of Mechanical Engineering, 53*(13), 29. https://doi.org/10.3901/jme.2017.13.029
8. Piskarev, E., Shintake, J., Ramachandran, V., Baugh, N., Dickey, M. D., & Floreano, D. (2020). Lighter and stronger: Cofabricated electrodes and variable stiffness elements in dielectric actuators. *Advanced Intelligent Systems, 2*(10), 2000069.
9. Zaidi, S., Maselli, M., Laschi, C., & Cianchetti, M. (2021). Actuation technologies for soft robot grippers and manipulators: A review. *Current Robotics Reports, 2*(3), 355–369. https://doi.org/10.1007/s43154-021-00054-5
10. Lee, K.-S., Kim, Y., & Park, H.-S. (2024). Shape memory alloy-based reactive tubular (SMART) brake for compact and energy-efficient wearable robot design. *ACS Applied Materials & Interfaces, 16*(7), 8974–8983. https://doi.org/10.1021/acsami.3c15179
11. Then Mozhi, G., Dhanalakshmi, K., & Choi, S.-B. (2023). Design and control of monolithic compliant gripper using shape memory alloy wires. *Sensors, 23*(4), 2052. https://doi.org/10.3390/s23042052
12. Lee, J.-H., Chung, Y. S., & Rodrigue, H. (2019). Long shape memory alloy tendon-based soft robotic actuators and implementation as a soft gripper. *Scientific Reports, 9*(1), 11251. https://doi.org/10.1038/s41598-019-47794-1
13. Safaei, K., et al. (2021). Additive manufacturing of NiTi shape memory alloy for biomedical applications: Review of the LPBF process ecosystem. *JOM, 73*(12), 3771–3786. https://doi.org/10.1007/s11837-021-04937-y
14. Xu, L., et al. (2017). Bio-inspired annelid robot: A dielectric elastomer actuated soft robot. *Bioinspiration & Biomimetics, 12*(2), 025003. https://doi.org/10.1088/1748-3190/aa50a5

15. Luo, H., et al. (2023). Review of dielectric elastomer based application for soft robotics. *Highlights in Science Engineering and Technology, 42*, 172–178. https://doi.org/10.54097/hset.v42i.7092
16. Hughes, J., Culha, U., Giardina, F., Guenther, F., Rosendo, A., & Iida, F. (2016). Soft manipulators and grippers: A review. *Frontiers in Robotics and AI, 3*, 223168.
17. Sut, D. J., & Sethuramalingam, P. (2023). Soft manipulator for soft robotic applications: A review. *Journal of Intelligent & Robotic Systems, 108*, 10.
18. Zhao, L., Jiang, Y., She, C., Chen, M., & Balkcom, D. (2024). SoftSnap: Rapid prototyping of untethered soft robots using snap-together modules.
19. Gu, H., Möckli, M., Ehmke, C., Kim, M., Wieland, M., Moser, S., Bechinger, C., Boehler, Q., & Nelson, B. J. (2023). Self-folding soft-robotic chains with reconfigurable shapes and functionalities. *Nature Communications, 14*(1), 1–12.
20. Toshimitsu, Y., Wong, K. W., Buchner, T., & Katzschmann, R. (2021). SoPrA: Fabrication & dynamical modeling of a scalable soft continuum robotic arm with integrated proprioceptive sensing. ArXiv.
21. Singh, A. V., Ansari, M. H., Dey, A. K., Laux, P., Samal, S. K., Malgaretti, P., Mohapatra, S. R., Busse, M., Suar, M., Tisato, V., & Gemmati, D. (2025). Bioinspired soft machines: Engineering nature's grace into future innovations. *Journal of Functional Biomaterials, 16*(5), 158.
22. Khosravi, H., Iannucci, S. M., & Li, S. (2021). Pneumatic soft actuators with kirigami skins. *Frontiers in Robotics and AI, 8*, 749051.
23. Walker, K. L., Partridge, A. J., Chen, H., Ramachandran, R. R., Stokes, A. A., Tadakuma, K., & Da Silva, L. C. (2024). A modular, tendon driven variable stiffness manipulator with internal routing for improved stability and increased payload capacity. ArXiv.
24. Onose, R., & Sawada, H. (2024). A ball-jointed tendon-driven continuum robot with multi-directional operability for grasping objects. *Robomech Journal, 11*, 4.
25. Li, H., Ford, C. J., Bianchi, M., Catalano, M. G., Psomopoulou, E., & Lepora, N. F. (2022). BRL/Pisa/IIT SoftHand: A low-cost, 3d-printed, underactuated, tendon-driven hand with soft and adaptive synergies. ArXiv.
26. Rafsanjani, A., et al. (2018). Kirigami skins make a simple soft actuator crawl. *Science Robotic, 3*, eaar7555.
27. Mawah, S. C., & Park, Y. (2023). Tendon-driven variable-stiffness pneumatic soft gripper robot. *Robotics, 12*(5), 128.
28. Mishra, A. K., Del Dottore, E., Sadeghi, A., Mondini, A., & Mazzolai, B. (2017). SIMBA: Tendon-driven modular continuum arm with soft reconfigurable gripper. *Frontiers in Robotics and AI, 4*, 223298.
29. Uthayasooriyan, A., Vanegas, F., Jalali, A., Digumarti, K. M., Janabi-Sharifi, F., & Gonzalez, F. (2024). Tendon-driven continuum robots for aerial manipulation—A survey of fabrication methods. *Drones, 8*, 269.
30. Luo, K. (2023). Modeling of continuum robots: A review. *Journal of Physics: Conference Series, 2634*, 012029. https://doi.org/10.1088/1742-6596/2634/1/012029
31. Wang, X., & Rojas, N. (2024). Cosserat rod modeling and validation for a soft continuum robot with self-controllable variable curvature. ArXiv.
32. Renda, F., Boyer, F., Dias, J., & Seneviratne, L. (2017). Discrete Cosserat approach for multi-section soft robots dynamics. ArXiv.
33. Isbister, A., Bailey, N. Y., & Georgilas, I. (2021). An integrated kinematic modeling and experimental approach for an active endoscope. *Frontiers in Robotics and AI, 8*, 667205.
34. Dubied, M., Michelis, M., Spielberg, A., & Katzschmann, R. (2022). Sim-to-real for soft robots using differentiable FEM: Recipes for meshing, damping, and actuation. ArXiv.
35. Guo, Y., Qin, Q., Han, Z., Plamthottam, R., Possinger, M., & Pei, Q. (2023). Dielectric elastomer artificial muscle materials advancement and soft robotic applications. *SmartMat, 4*(4), e1203.
36. Wang, Y., Wang, Y., Mushtaq, R. T., & Wei, Q. (2023). Advancements in soft robotics: A comprehensive review on actuation methods, materials, and applications. *Polymers, 16*(8), 1087.

37. Hao, Y., Zhang, S., Fang, B., et al. (2022). A review of smart materials for the boost of soft actuators, soft sensors, and robotics applications. *Chinese Journal of Mechanical Engineering, 35*, 37.
38. Sato, Y., Yuhn, C., Kobayashi, H., Kawamoto, A., & Nomura, T. (2024). Computational co-design of structure and feedback controller for locomoting soft robots. ArXiv.
39. Hunt, J., & Lee, H. (2021). Optimizing the rigid or compliant behavior of a novel parallel-actuated architecture for exoskeleton robot applications. *Frontiers in Robotics and AI, 8*, 596958.
40. Kallioras, N. A., & Lagaros, N. D. (2020). MLGen: Generative design framework based on machine learning and topology optimization. *Applied Sciences, 11*(24), 12044.
41. Wang, T., Ma, P., Spielberg, A. E., Xian, Z., Zhang, H., Tenenbaum, J. B., Rus, D., & Gan, C. (2023). SoftZoo: A soft robot co-design benchmark for locomotion in diverse environments. ArXiv.
42. Mertan, A., & Cheney, N. (2023). Modular controllers facilitate the co-optimization of morphology and control in soft robots. ArXiv.
43. Roshanfar, M., Dargahi, J., & Hooshiar, A. (2024). Design optimization of a hybrid-driven soft surgical robot with biomimetic constraints. *Biomimetics, 9*(1), 59.
44. Jia, J., Sun, X., Liu, T., Tang, J., Wang, J., & Hu, X. (2023). Structural optimization design of dual robot gripper unloading device based on intelligent optimization algorithms and generative design. *Sensors (Basel, Switzerland), 23*(19), 8298.
45. Spillman, W. B., Jr., Sirkis, J. S., & Gardiner, P. T. (1996). Smart materials and structures: What are they? *Smart Materials and Structures, 5*(3), 247–254.
46. Case, J. C., White, E. L., & Kramer, R. K. (2015). Soft material characterization for robotic applications. *Soft Robotics, 2*(2), 80–87.
47. Wallin, T. J., Pikul, J., & Shepherd, R. F. (2018). 3D printing of soft robotic systems. *Nature Reviews Materials, 3*(6), 84–100.
48. Kwon, H. J., Osada, Y., & Gong, J. P. (2006). Polyelectrolyte gels-fundamentals and applications. *Polymer Journal, 38*(12), 1211–1219.
49. Koetting, M. C., Peters, J. T., Steichen, S. D., & Peppas, N. A. (2015). Stimulus-responsive hydrogels: Theory, modern advances, and applications. *Materials Science and Engineering: R: Reports, 93*, 1–49.
50. Willner, I. (2017). Stimuli-controlled hydrogels and their applications. *Accounts of Chemical Research, 50*(4), 657–658.
51. Kim, D., Lee, H. S., & Yoon, J. (2016). Highly bendable bilayer-type photo-actuators comprising of reduced graphene oxide dispersed in hydrogels. *Scientific Reports, 6*, 20921.
52. Osada, Y., Okuzaki, H., & Hori, H. (1992). A polymer gel with electrically driven motility. *Nature, 355*(6357), 242–244.
53. Liang, S., Weng, L., Tan, S., Xu, J., Zhang, X., & Zhang, L. (2007). Field-driven gel actuator with versatile long-range locomotion in air. *Applied Physics Letters, 90*(15), 153506–153508.
54. Uddin, M. Z., Watanabe, M., Shirai, H., & Hirai, T. (2003). Efects of plasticizers on novel electromechanical actuations with different poly (vinyl chloride) gels. *Journal of Polymer Science Part B: Polymer Physics, 41*(18), 2119–2127.
55. Kaneto, K. (2016). Research trends of soft actuators based on electroactive polymers and conducting polymers. *Journal of Physics: Conference Series, 704*(1), 012004–012012.
56. Baughman, R. H., Cui, C., Zakhidov, A. A., Iqbal, Z., Barisci, J. N., Spinks, G. M., et al. (1999). Carbon nanotube actuators. *Science, 284*(5418), 1340–1344.
57. Pelrine, R., Kornbluh, R., Pei, Q., & Joseph, J. (2000). Highspeed electrically actuated elastomers with strain greater than 100%. *Science, 287*(5454), 836–839.
58. Leng, J., Lan, X., Liu, Y., & Du, S. (2011). Shape-memory polymers and their composites: Stimulus methods and applications. *Progress in Materials Science, 56*(7), 1077–1135.
59. Kim, J.-H., Shim, B. S., Kim, H. S., Lee, Y.-J., Min, S.-K., Jang, D., et al. (2015). Review of nanocellulose for sustainable future materials. *International Journal of Precision Engineering and Manufacturing-Green Technology, 2*(2), 197–213.

Drive and Perception Systems of Flexible Manipulators

1 Flexible Robotic Arm Electric-Gas-Liquid Drive Systems: Technical Challenges and Frontier Directions

The rapid advancement of flexible robotic arm technology has revolutionized operations across critical domains, from minimally invasive surgical interventions and precision automotive assembly to extreme-environment tasks like satellite maintenance and deep-sea exploration. Central to this transformation are electric, pneumatic, and hydraulic drive systems—each embodying unique trade-offs between precision, force density, flexibility, and environmental adaptability. Electric actuators, for instance, deliver sub-micron positioning for surgical robots, while pneumatic systems excel in safe human-robot collaboration, and hydraulic drives dominate heavy-load scenarios. Yet, as detailed in preceding analyses of drive system designs and applications, the path toward robust, scalable, and autonomous deployment of flexible robotic arms remains hindered by profound technical barriers.

This chapter shifts focus from established drive system architectures and application successes to the unsolved challenges that define the frontier of this field. While earlier discussions outlined the principles, performance metrics, and use cases of electric, pneumatic, and hydraulic drives, we now confront the critical limitations that compromise their reliability, efficiency, and broader adoption. Addressing these challenges is not merely an engineering exercise; it is an imperative for unlocking the next generation of flexible robotic arms capable of operating safely and effectively in dynamic, real-world settings—where variable loads, environmental perturbations, and unstructured workflows prevail.

The significance of this discussion extends beyond immediate technical hurdles. Breakthroughs in drive material durability, nonlinear control, energy efficiency, and system integration will catalyze innovations in personalized healthcare (e.g., implantable robotic devices), sustainable manufacturing (e.g., low-energy assembly

J. Qu, *Robotic Flexible Manipulation and Intelligent Sensing Technologies*,
https://doi.org/10.1007/978-981-95-6232-9_3

lines), and disaster response (e.g., ruggedized rescue robots). Conversely, unresolved challenges risk confining flexible robotic arms to controlled laboratory environments or niche applications, limiting their transformative potential. The development of lead-free piezoelectric ceramics (such as KNN-based materials) has also improved the thermal stability of the actuator in vacuum environments [1].

This chapter systematically dissects these bottlenecks across four interconnected domains: *drive material performance and durability, nonlinear control and precision regulation, energy efficiency and sustainable operation*, and *system integration and environmental adaptability*. Our goal is twofold: to provide researchers and engineers with a clear road map of critical gaps that demand attention and to inspire collaborative, interdisciplinary efforts to overcome them (Table 1).

1.1 Drive Material Performance and Durability

The performance of flexible robotic arms hinges on advanced materials that balance flexibility, strength, and responsiveness—yet these materials face intrinsic limitations that propagate through every layer of drive system design. Multifunctional material requirements create fundamental trade-offs: high elasticity for compliant deformation (e.g., 300% strain in dielectric elastomers) often sacrifices tensile strength, while electrical conductivity (critical for self-sensing SMA actuators) degrades under cyclic loading. Even state-of-the-art composites, such as nanoparticle-reinforced silicones or liquid metal-embedded elastomers, struggle to simultaneously deliver biocompatibility, self-healing capability, and environmental resilience. For example, liquid-metal circuits in soft pneumatic actuators offer exceptional deformability but suffer from motion-induced signal drift, forcing designers to compromise between sensing stability and flexibility. Equipped with high-fidelity tactile sensors, the system demonstrates contact pressure resolution of 0.1 N, significantly enhancing charging reliability [2].

Durability under operational stress emerges as a critical frontier. Repeated large-strain deformation induces microstructural fatigue in key drive materials: chain

Table 1 Driver performance comparison table

Driver type	Force density (kN/kg)	Velocity range (m/s)	Energy efficiency (%)	Positioning expertise linear measure	Temperature adaptability range (°C)
Fluid drive	15–22	0.1–0.5	40–60	± 0.5 mm rice	−20 to 80
Pneumatic drive	2–8	0.5–1.2	30–40	±2 mm	5–50
Servo electric	3–12	0.05–0.3	75–90	±0.01 mm rice	−10 to 40
Piezoelectric drive	0.1–0.5	0.001–0.1	85–95	±1 nm	10–35
Gas-electricity mix	5–15	0.2–0.8	50–70	± 0.1 mm rice	−10 to 60

scission in polymer-based pneumatic actuators, delamination in stretchable electrodes of electric systems, and void nucleation in hydraulic chamber seals. Environmental aggressors accelerate degradation: UV radiation embrittles silicones in outdoor applications, body fluids swell hydrogels in medical robots, and industrial chemicals plasticize polyurethanes in manufacturing settings. While self-healing materials (e.g., polysiloxane-polyether elastomers) partially mitigate damage—repairing 5 mm-wide incisions within 24 h at room temperature—their healing cycles are finite, and activation often requires heat or light inaccessible in confined spaces (e.g., endoscopic channels). Consequently, field-deployable systems like disaster-rescue robotic arms face unpredictable "lifespan cliffs," where a single cut or abrasion disables critical drive segments.

Within an 8% strain range, the resistance can change at a rate of up to 12%, as detailed in reference [3]. Material limitations also compound control challenges. Hysteresis in shape memory alloy (SMA) actuators—caused by phase-transition irreversibility—introduces position errors [4] of up to 12% without real-time sensor feedback, forcing reliance on fragile strain gauges. Creep in viscoelastic tendons of electric drive systems causes grasp-force relaxation in soft grippers during prolonged holds, risking object drops in assembly lines. Thermal management exacerbates these issues: Joule heating from embedded electronics in electric actuators, combined with phase-transition heat in SMAs, can trigger thermal runaway, degrading adjacent sensors or structural integrity. This interdependence demands material-control co-design approaches currently hindered by disciplinary silos: materials scientists optimize strain tolerance, while control engineers battle hysteresis-induced instability.

Emerging solutions include bio-inspired material architectures [5], such as MIT's 3D-printed hydrogel muscles with micrometer-scale alignment grooves (enabling ±180° coordinated contraction) or Northwestern University's "ossified muscles" (gradient stiffness rubber layers that boost localized stiffness to 2.3 MPa while retaining flexibility). Magnetic-responsive phase-change materials also show promise, enabling rapid solid-to-liquid transitions within 5 s for complex deformations. However, standardizing material testing protocols—e.g., cyclic fatigue benchmarks across 10^4+ cycles—and scaling manufacturing of these advanced materials remain critical to accelerating progress. As exemplified in satellite maintenance hydraulic systems, where rapeseed-oil media extends service life by 15% but fails at 20 MPa due to rotor wear, overcoming material barriers will unlock more durable, versatile flexible robotic arms.

1.2 Nonlinear Control and Precision Regulation

The transition from rigid to flexible robotic arms fundamentally disrupts traditional control paradigms. Where conventional robots leverage deterministic kinematics and linear dynamics, flexible drive systems—characterized by distributed deformation, underactuation, and nonlinearities (hysteresis, creep, viscoelasticity)—demand controllers that reconcile adaptability with precision in the face of intrinsic uncertainty. Model-based approaches, while theoretically grounded, falter when confronted with imperfect representations of drive system behavior. Inaccuracies in simulating SMA phase transitions, pneumatic pressure-volume hysteresis, or hydraulic fluid leakage propagate directly into control loops, degrading performance.

Model predictive control (MPC), for instance, relies on real-time deformation forecasts to optimize trajectories for pneumatic grippers in automotive assembly [6]. Yet computational delays (exceeding 8 ms for 12-layer neural network models) or unmodeled disturbances—such as air leakage in soft pneumatic chambers or tendon friction in electric drives—can cause oscillations or task failure, e.g., damaging delicate automotive components or misaligning surgical tools. This fragility amplifies when pursuing high-precision regulation: achieving millimeter-scale tip positioning in electric endoscopic robots or sub-Newton force control in pneumatic food grippers requires compensating for dynamic coupling between drive segments and environmental compliance. While impedance control frameworks offer partial solutions, they struggle with the dual challenge of large, rapid deformations (e.g., 400% strain in pneumatic actuators) and fine force modulation—especially when sensor feedback is sparse or noisy (e.g., drift in liquid-metal strain sensors).

Learning-based methods—reinforcement learning (RL) and imitation learning (IL) [7]—have emerged as alternatives, promising adaptability through data-driven policy optimization. A recent RL algorithm, trained on 1.03 million steps, achieved 0.123 mm trajectory tracking accuracy for pneumatic surgical robots—an 84.95% improvement over traditional PID control. However, these approaches face steep barriers in flexible drive systems. Sample inefficiency plagues RL: physical experiments on deformable structures are slow and risk damaging delicate materials (e.g., hydrogel actuators). Sim-to-real transfer remains problematic due to the "reality gap"—simulated models of pneumatic softness or hydraulic fluid dynamics fail to capture real-world variability (e.g., temperature-induced viscosity changes in hydraulic oil [8]). More critically, safety is compromised: black-box neural network policies lack interpretability, may ignore force constraints during exploration, and fail catastrophically under novel conditions—a prohibitive risk in human-in-the-loop scenarios like rehabilitation robotics.

Control challenges extend to motion planning and autonomy. Planning collision-free paths for hydraulic robotic arms in satellite maintenance involves navigating high-dimensional configuration spaces with complex constraints (e.g., zero-gravity dynamics and confined satellite bays). While sampling-based planners handle obstacles, they disregard deformation energy minimization or dynamic stability, leading to inefficient or unstable motions. Enabling true autonomy—where systems self-adapt to drive wear (e.g., SMA strain degradation) or payload shifts—demands online re-planning capabilities currently hindered by computational latency and incomplete world models. Even teleoperation, a stopgap for complex tasks, grapples with haptic transparency: transmitting realistic force feedback through flexible hydraulic interfaces requires low-latency, distortion-free control loops rarely achievable with current sensor-controller co-design.

Promising pathways include hybrid neuro-symbolic architectures that blend RL's adaptability with physics-based safety filters (e.g., enforcing force limits in surgical robots [9]). Online model identification—updating controller parameters in real-time to account for SMA hysteresis or hydraulic leakage—also shows potential, as demonstrated in underwater robotic arms that maintain ±0.2 mm positioning accuracy despite temperature fluctuations. Additionally, neuromorphic control systems, built on chips like Intel Loihi, aim to achieve autonomous decision-making under ultra-low energy budgets (1 mJ), critical for portable flexible robots. Overcoming these hurdles demands tighter integration of control theory with material science and sensing technology—e.g., developing self-sensing materials that reduce reliance on external sensors and lower computational overhead.

1.3 Energy Efficiency and Sustainable Operation

Energy efficiency and sustainability have become defining constraints for flexible robotic arm deployment, particularly in portable (e.g., wearable rehabilitation devices [10]) or long-duration (e.g., space maintenance) applications. Current drive systems exhibit significant inefficiencies rooted in their working principles: hydraulic systems suffer from fluid leakage and pump friction (efficiency <30%), pneumatic systems waste energy on gas compression and leakage (efficiency 30–40%), and even high-efficiency servo electric systems (75–90%) struggle with standby power consumption during idle periods in assembly lines. These inefficiencies limit operational lifetimes—e.g., battery-powered electric endoscopic robots require frequent recharging—or increase operational costs, hindering adoption in resource-constrained settings.

The energy challenges of each drive type are compounded by their unique characteristics. Pneumatic systems [11], while lightweight and safe, face inherent losses from isothermal gas expansion: up to 40% of input energy is dissipated as heat in soft actuators, and porous elastic materials exacerbate leakage (rates three times higher than traditional rigid cylinders). Hydraulic systems, though high in force density (15–22 kN/kg), lose energy to internal leakage (5.8% flow deviation at 3 MPa) and viscous friction, which increases exponentially at low temperatures (ISO VG43 oil viscosity triples at −15 °C, reducing response speed by 40%). Electric systems, while precise, suffer from power losses in high-voltage circuits for piezoelectric actuators (45% of energy converted to heat at 1 kHz) and static power draw in stepper motors (20–40% of rated power during position holding).

Sustainable operation further demands addressing material lifecycle and energy recovery. Many drive materials—e.g., piezoelectric ceramics or hydraulic fluids [12]—are non-biodegradable, creating disposal challenges for medical or space applications. Self-healing materials offer a partial solution (e.g., 92% tensile strength recovery in polysiloxane elastomers) but remain limited by fatigue life (<10^4 cycles for current formulations). Energy recovery systems, such as ABB's YuMi robotic arm with kinetic energy recovery (KERS) via piezoelectric composites, reduce net consumption by 22% but add complexity and weight. Ionic fuel-driven hydrogel actuators, which achieve 85% mechanical-to-chemical energy conversion efficiency, represent a promising sustainable alternative—but their output force (<50 N) currently limits applicability to micro-robotic tasks.

Hybrid drive systems have emerged as a strategy to balance efficiency and performance [13]. Gas-electric hybrid grippers, for example, use pneumatic modules for rapid positioning (0.8 m/s over 500 mm) and electric modules for ±0.05 mm precision, reducing energy consumption by 35% compared to pure-pneumatic systems. Hydraulic-electric hybrids for construction machinery switch to electric mode during silent relocation, cutting fuel use by 38% in 10-ton lifting operations. However, these systems introduce new challenges: multi-domain energy management requires real-time optimization of power distribution (e.g., balancing battery SOC and hydraulic tank pressure), and component integration increases complexity and cost.

Future progress will depend on three key areas:

1. Material-level innovations, such as low-loss dielectric elastomers or self-lubricating hydraulic seals, to reduce intrinsic energy dissipation.
2. System-level optimization, including MPC-based energy allocation to minimize waste in hybrid drives.
3. Sustainable energy sources, such as integrating friction nanogenerators or biofuel cells to create self-powered robotic arms (targeting >50% energy conversion efficiency). As exemplified by ionic hydrogel actuators that operate continuously for 2 h per charge, overcoming energy barriers will enable flexible robotic arms to operate independently in remote or harsh environments.

1.4 System Integration and Environmental Adaptability

The reliable operation of flexible robotic arms depends on seamless integration of drive systems, sensors, and control architectures—yet this integration is complicated by conflicting requirements: high-density sensor arrays (for precise state estimation) add weight and power draw, while lightweight drive materials (for flexibility) limit sensor embedding [14]. Compounding this is the need for environmental adaptability: robotic arms must operate in diverse conditions, from the sterile, temperature-controlled environment of operating rooms to the extreme temperatures (−20 to 80 °C) of construction sites or the vacuum of space. These challenges create bottlenecks that limit deployment beyond controlled settings.

Sensor-drive integration faces fundamental trade-offs. Distributed sensing networks—critical for monitoring deformation in flexible arms—must balance spatial resolution, mechanical robustness, and signal stability. Optical fiber sensors (FBGs) offer millimeter-scale strain mapping but fracture at large elongations (>8%); liquid-metal circuits tolerate 400% strain but suffer from motion-induced drift. This forces designers into compromises: surgical robots may use FBGs for precision but risk sensor failure during large deformations, while industrial grippers rely on less accurate pressure sensors to ensure durability. Multimodal sensor fusion—combining tactile arrays, inertial units, and cameras—exacerbates integration challenges: differing bandwidths (kHz for strain gauges vs. 30 Hz for vision) create spatiotemporal misalignment, and sensor shift relative to deformable substrates introduces calibration errors. Bayesian filters or deep fusion networks struggle to resolve these conflicts, particularly in occlusion-prone environments (e.g., cluttered surgical fields or pipe interiors) [15].

Environmental adaptability exposes further vulnerabilities. Hydraulic systems are sensitive to fluid properties: temperature fluctuations alter viscosity (reducing efficiency by 40% at −15 °C), while water-based media cause corrosion in uncoated components. Pneumatic systems are prone to humidity-induced failures: each 5 °C rise in dew point increases solenoid valve failure rates by 17%. Electric systems face electromagnetic interference (EMI) in industrial settings (disrupting servo motor control) and dielectric breakdown in high-humidity environments (e.g., 80% relative humidity for piezoelectric actuators). Even advanced materials struggle: SMA actuators lose 30% of strain capacity after 10^4 cycles in corrosive industrial atmospheres, and soft pneumatic elastomers degrade 20% faster under UV exposure [16].

System integration also poses practical challenges. Miniaturization of hydraulic drives [17]—critical for endoscopic or satellite applications—reduces leakage tolerance: micro-hydraulic cylinders (5 mm diameter) exhibit leakage rates >0.1 mL/min due to microscale sealing defects. Electric drive integration requires balancing high-precision components (e.g., 17-bit encoders for servo motors) with flexible structures, often leading to bulky or fragile assemblies. Hybrid systems, while performance-optimized, require complex interfaces: electro-hydraulic units (EHUs) need standardized ISO 3020/3022 interfaces to compatibility with traditional cylinders, and gas-electric hybrids demand synchronization between pneumatic valves and electric motors.

Emerging solutions focus on modular, adaptive integration. Pressure-compensated sealing mechanisms for hydraulic arms enable deep-sea operation by mitigating pressure-induced leakage. Self-diagnostic algorithms detect sensor drift or drive wear (e.g., identifying hydraulic pump wear via 3–5 kHz vibration signatures) but cannot compensate for irreversible degradation, requiring periodic recalibration [18]. Three-dimensional printing enables customized integration: rapid liquid-printed silicone actuators integrate embedded sensors and fluid channels in a single step, matching the performance of injection-molded parts while reducing assembly complexity. Standardization—e.g., defining test protocols for environmental durability or sensor calibration—also accelerates progress, as seen in the development of ISO-compliant pneumatic grippers [19] for automotive assembly.

Overcoming integration and adaptability barriers will require interdisciplinary collaboration: materials scientists developing environment-resistant, sensor-embedded materials; mechanical engineers designing modular drive architectures; and control engineers optimizing multi-domain coordination. As exemplified by NASA's remote hydraulic manipulators, which operate reliably in zero-gravity via vibration-monitored wear prediction, solving these challenges will unlock flexible robotic arms capable of consistent performance across the full range of real-world environments [20].

2 Tactile Perception

Tactile perception represents a cornerstone of intelligent sensing in flexible manipulation technologies, enabling robotic systems to interact with unstructured environments in a manner that mimics human touch.

Unlike rigid manipulators, which often rely on predefined force limits and visual cues alone, flexible manipulation systems such as soft robotic arms, continuum manipulators, and compliant end-effectors demand nuanced haptic feedback to grasp, manipulate, and release objects without causing damage [24].

This feedback encompasses a spectrum of sensory modalities, including pressure distribution, shear forces, texture recognition, slip detection, and even thermal variations. The integration of tactile sensors into these systems transforms passive compliance into active intelligence, allowing for adaptive control strategies that respond to object properties like fragility, shape, and surface roughness [25, 26].

The evolution of tactile perception has been driven by the need to bridge the gap between mechanical dexterity and sensory acuity. Early flexible manipulators focused primarily on kinematic design and actuation mechanisms, such as pneumatic bellows or cable-driven linkages, to achieve gross motion. However, without tactile sensing, these systems struggled with precision tasks, such as handling delicate electronics or biological tissues, where excessive force could lead to failure. Recent advancements have introduced a diverse array of sensor technologies, leveraging novel materials and transduction principles to achieve high sensitivity, durability, and multimodality. These sensors not only detect contact forces but also facilitate closed-loop control, enabling manipulators to adjust grip strategies in real-time [27, 28].

In this section, we categorize tactile sensors for flexible manipulation into three broad groups: traditional tactile sensors, which form the foundational backbone due to their maturity and reliability; emerging tactile sensors, which push the boundaries of resolution and integration; and other advanced tactile sensors, representing cutting-edge innovations. Traditional sensors, including capacitive, piezoresistive, piezoelectric, and fiber Bragg grating (FBG) types, excel in static and dynamic force measurement with proven robustness. Emerging variants, such as vision-based and triboelectric sensors, offer enhanced spatial resolution and energy efficiency, ideal for complex environments. Finally, other sensors explore hybrid or bio-inspired approaches to address unmet needs like extreme durability or multifunctionality. By examining these categories, we highlight their principles, structures, performance metrics, and applications in flexible manipulation, while underscoring ongoing challenges such as hysteresis, noise susceptibility, and integration with control algorithms.

Traditional tactile sensors have long served as the workhorses of flexible manipulation, providing reliable transduction of mechanical stimuli into electrical or optical signals. Their designs emphasize simplicity, cost-effectiveness, and compatibility with soft materials like polydimethylsiloxane (PDMS) or silicone elastomers, which are ubiquitous in compliant manipulators. These sensors typically operate on principles rooted in electromechanical or optical effects, offering trade-offs between sensitivity, response time, and environmental resilience. Below, we delve into the key subtypes, drawing parallels to their deployment in continuum arms or soft-fingered end-effectors [29–31].

2.1 Capacitive Tactile Sensors

Capacitive tactile sensors detect deformation through changes in capacitance, a principle that exploits the geometry of a parallel-plate capacitor embedded within a flexible substrate. The capacitance C is governed by the equation:

$$C = \frac{\epsilon_0 \epsilon_r S}{d}$$

where ϵ0 is the permittivity of free space, ϵr is the relative permittivity of the dielectric, S is the overlapping electrode area, and d is the inter-electrode gap. External forces alter S (for shear detection) or d (for normal pressure), enabling precise mapping of contact profiles.

Structurally, these sensors feature a sandwich configuration: two compliant electrodes (e.g., carbon nanotubes (CNTs), graphene, or indium tin oxide (ITO)) sandwiching a low-modulus dielectric like Ecoflex silicone or PDMS. Microstructured dielectrics—such as pyramids, domes, or air gaps—enhance sensitivity by amplifying geometric changes, achieving resolutions down to 3 Pa and ranges up to 1.7 MPa. For instance, a gradient micro-dome architecture (GDA) integrates hierarchical PDMS layers to mitigate viscoelastic damping, yielding a sensitivity of 0.065 kPa^{-1} with a 100 ms response time.

In flexible manipulation, capacitive sensors shine in applications requiring static force feedback, such as a soft continuum arm grasping irregular fruits. Their low power consumption (often <1 mW per taxel) and high spatial resolution (up to 1 kHz bandwidth) facilitate array configurations for distributed sensing along a manipulator's length. A notable example is their use in a cable-driven soft finger, where interdigitated electrodes detect both normal and tangential forces, enabling slip prevention during tele-operated assembly tasks.

Merits include excellent dynamic range and noise immunity in controlled settings, but demerits such as electromagnetic interference susceptibility and hysteresis from dielectric creep limit their use in high-vibration scenarios. Cost-effective fabrication via screen-printing keeps them accessible, with arrays scalable to hundreds of taxels for full-arm coverage.

2.2 Piezoresistive Tactile Sensors

Piezoresistive sensors transduce mechanical strain into resistance variations, leveraging the material's intrinsic sensitivity to stress-induced carrier mobility changes. The relative resistance change is approximated as:

$$C = \frac{\epsilon_0 \epsilon_r S}{d}$$

where ν is Poisson's ratio, π is the piezoresistive coefficient, E is Young's modulus, and ϵ is strain. This effect arises from geometric deformation (elongation reducing cross-section) or resistivity shifts in conductive fillers under compression.

Common structures employ a conductive composite matrix (e.g., CNTs, metal nanowires, or MXene embedded in PDMS or eco-elastomers) patterned into serpentine traces or interlocked microstructures. Sensitivity reaches 133 kPa^{-1} for ultra-low pressures (0.8 Pa), with ranges extending to 0.8 MPa. A fabric-based variant, using piezoresistive yarns in a multilayer weave, detects normal forces up to 500 kPa with 0.1 N resolution, ideal for wearable manipulator interfaces.

For flexible manipulation, these sensors enable in situ monitoring of bending and stretching in pneumatic actuators. In a bio-inspired octopus-like arm, eutectic gallium-indium (EGaIn) channels provide multimodal feedback on elongation and torsion, supporting dexterous object reorientation. Their high signal-to-noise ratio (SNR) minimizes crosstalk in dense arrays, though hysteresis (up to 10% drift) and moderate frequency response (40–200 Hz) restrict them to quasi-static tasks like tissue palpation.

Advantages encompass simple integration and robustness to shear, but energy demands from resistive readout circuits pose challenges. Nonetheless, their low cost and ease of 3D printing make them staples in prototype manipulators.

2.3 Piezoelectric Tactile Sensors

Piezoelectric sensors harness the direct piezoelectric effect, generating charge from material polarization under stress:

$$D_i = d_{ijk} \sigma_{jk}$$

where Di is charge density, dijk is the piezoelectric coefficient, and σjk is applied stress. Active variants exploit resonant frequency shifts for enhanced precision.

These sensors typically comprise thin films of polyvinylidene fluoride (PVDF), lead zirconate titanate (PZT), or zinc oxide (ZnO) laminated between flexible electrodes on elastomeric substrates. PVDF's flexibility ($d_{33} \approx$ 20–30 pC/N) suits soft manipulators, while PZT hybrids offer higher sensitivity (0.005 Pa threshold).

Response times as low as 0.1 ms and bandwidths up to 1 kHz enable vibration isolation in dynamic grasping.

In manipulation contexts, piezoelectric arrays detect slip in underwater robotic arms, where silicone-embedded nanowires sense normal forces up to 750 kPa. A wrist-mounted PZT square array, for example, maps multi-axis strains during collaborative assembly, feeding data to impedance controllers for compliant interaction.

Their high-frequency response excels for transient events like impact detection, and self-powering via energy harvesting reduces wiring needs. However, the inability to sense static forces and temperature sensitivity (up to 0.1% shift/°C) necessitate compensation circuits. Low cost and minimal bulkiness position them as enablers for lightweight, untethered manipulators.

2.4 *Fiber Bragg Grating Tactile Sensors*

FBG sensors utilize optical wavelength shifts in inscribed fiber gratings to measure strain:

$$\lambda_B = 2n_{eff}\Lambda$$

where λB is the Bragg wavelength, neff is the effective refractive index, and Λ is the grating period. Strain alters Λ, yielding shifts proportional to force (sensitivity ~2.2 mN^{-1}).

Embedded in silicone sheaths along a manipulator's backbone, FBG arrays provide three-axis force mapping with 0.01 N resolution over 4 N ranges. Their immunity to electromagnetic interference (EMI) and tolerance to extremes (−196 °C to 1000 °C) suit harsh environments, such as nuclear decommissioning arms.

In flexible systems, FBGs enable curvature sensing in continuum robots, detecting slip and roughness via wavelength demodulation at 100 Hz. A fiber-optic integrated arm, for instance, uses multiplexed gratings for distributed feedback, enhancing precision in tele-surgery.

High SNR and vibration resistance are key strengths, but complex interrogators inflate costs. Despite this, their longevity (millions of cycles) justifies use in high-stakes manipulation.

2.5 *Emerging Tactile Sensors*

Emerging sensors leverage interdisciplinary advances in optics, nanomaterials, and electrostatics to overcome traditional limitations, offering unprecedented resolution and multifunctionality for next-generation flexible manipulators [32–34].

1. Vision-Based Tactile Sensors

Vision-based sensors fuse optical imaging with elastomeric interfaces, capturing deformation via embedded markers or textures viewed by internal cameras. A gelatinous skin with LED-illuminated markers, for example, yields 128 × 128 pixel maps at 40 fps, detecting shapes and slips with 3 N range.

In manipulation, GelSight-inspired systems on Fin Ray fingers visualize sliding directions, enabling adaptive grasping of deformable objects like fabrics. High dynamic range and frequency (up to 40 Hz) support real-time pose estimation, though light source dependency increases complexity.

2. Triboelectric Tactile Sensors

Triboelectric nanogenerators (TENGs) convert contact electrification into voltage via layered tribomaterials (e.g., PTFE and Cu on PET substrates). Sensitivity hits 0.06 kPa^{-1}, with self-powering for curvature (up to 120 mm) and object identification.

Deployed in soft grippers (adapted here to manipulators), they detect 1 kPa pressures at 100 Hz, harvesting energy for wireless operation. Low cost and efficiency shine, but durability under cyclic stress remains a hurdle.

3. Other Advanced Tactile Sensors

Beyond core categories, hybrid and bio-inspired sensors address niche demands. Liquid-metal microfluidics, for instance, embed conductive fluids in microchannels for multi-axis strain detection, ideal for hyperelastic arms. Magnetic sensors using Hall-effect arrays sense proximity and force without physical contact, suiting non-invasive palpation. Biomimetic e-skins with iontronic hydrogels fuse ionic conduction for humidity-robust perception. These innovations, often combining principles (e.g., piezo-tribo hybrids), promise multimodal arrays for autonomous manipulation, though scalability and data fusion pose future challenges [35–37].

3 Visual Perception

3.1 Importance of Visual Perception in Flexible Manipulators

As a core component of the intelligent control system of flexible manipulators, visual perception provides an intuitive understanding of the environment and enables precise localization. Compared with conventional rigid manipulators, flexible manipulators exhibit inherent softness and complex deformation behaviors, which significantly increase their reliance on visual feedback. The visual system must not only perceive the position, shape, and motion state of target objects in the external environment but also monitor the manipulator's own morphological changes in real-time. This dual perception demand renders the visual system indispensable for the control of flexible manipulators.

Modern visual perception systems in flexible manipulators typically process multilayered information. At the macroscopic level, the system identifies and locates target objects within the workspace, analyzing their geometric features and surface attributes. At the mesoscopic level, it monitors the relative spatial relationship between the manipulator and the target, predicting contact points and the distribution of contact forces. At the microscopic level, it captures subtle deformations of the manipulator's end-effector and its deformation responses upon contact. Such multi-scale and multidimensional perception capabilities directly determine the operational precision and adaptability of flexible manipulators in complex environments.

Another crucial role of visual perception systems is to provide a foundation for autonomous learning and adaptation. Through visual feedback, the system can evaluate operational outcomes, analyze causes of failure, and progressively optimize control strategies. This vision-driven learning mechanism enables flexible manipulators to rapidly adapt to new tasks in unknown environments, thereby improving success rates and operational efficiency. Furthermore, visual systems play a critical role in safety monitoring by detecting potential collision risks and abnormal situations in real-time, ensuring safe operation in human–robot collaborative environments [21, 22].

3.2 Depth Estimation and Three-Dimensional Reconstruction

Depth estimation is a crucial capability of visual perception systems in flexible manipulators. In complex and unstructured environments, manipulators require accurate 3D spatial information to perform effective grasping and manipulation. Traditional monocular imaging can only provide 2D information, making it insufficient for precise spatial localization. Depth estimation allows the visual system to reconstruct the 3D structure of the environment, thereby offering essential support for manipulation tasks.

Current methods for depth estimation can be categorized into three main classes: binocular stereo vision, structured light, and deep learning-based monocular depth estimation. Binocular stereo vision estimates depth by calculating disparities between two viewpoints, providing a balance of accuracy and real-time performance. Structured light approaches project coded light patterns and analyze their deformations on object surfaces, yielding high-precision depth measurements but often with limited adaptability to complex environments. Deep learning-based monocular depth estimation, however, leverages large-scale datasets to predict pixel-level depth maps from single images, demonstrating strong generalization capability and adaptability in dynamic scenarios.

For flexible manipulators, 3D reconstruction based on depth estimation is especially critical for two tasks: (1) dynamic grasping of deformable objects and (2) accurate tracking of self-deformation. The reconstruction process involves generating point clouds or mesh models, which are then used for contact planning and force prediction. High-precision depth perception thus enhances the manipulator's ability to interact with objects of diverse shapes and materials [23–25].

3.3 Object Recognition and Semantic Understanding

Object recognition forms the basis of intelligent operation in flexible manipulators. Beyond mere detection of object position, modern systems must also understand the semantic attributes of objects, such as category, material, and functional properties. Semantic understanding enables the manipulator not only to grasp an object but also to determine how best to manipulate it.

Deep learning models, particularly convolutional neural networks (CNNs) and vision transformers (ViTs), play a central role in visual recognition tasks. By training on large-scale labeled datasets, these models can achieve high-accuracy classification and detection. Moreover, multimodal learning approaches combine vision with language models, allowing semantic reasoning about object functionality.

In practical applications, object recognition and semantic understanding are tightly coupled with task planning. For instance, when handling fragile objects, the manipulator must infer that excessive force may cause damage and thus choose a compliant grasping strategy. By integrating visual semantics into control loops, flexible manipulators can achieve higher levels of autonomy and adaptability in unstructured environments [26–28].

3.4 Motion Tracking and Trajectory Prediction

Real-time tracking of target motion is vital for tasks such as grasping moving objects, collaborative operations, and dynamic assembly. Motion tracking provides continuous position updates, while trajectory prediction estimates future motion paths, enabling the manipulator to proactively plan its actions.

Conventional approaches employ optical flow, feature tracking, and Kalman filters for estimating motion states. Recent advances have introduced deep learning-based trackers, which are capable of handling occlusions, illumination changes, and background clutter. These methods integrate temporal information to maintain robustness over long tracking sequences.

For flexible manipulators, trajectory prediction must account for the uncertainty introduced by both the target and the manipulator's own deformation. Probabilistic prediction models, such as recurrent neural networks (RNNs) and variational Bayesian methods, provide not only point estimates but also confidence intervals of predicted trajectories. This uncertainty-aware prediction improves the reliability of motion planning and reduces collision risk in dynamic environments [29, 30].

3.5 Feature Extraction and Representation

Feature extraction is a fundamental step in visual perception. It transforms raw image data into structured, high-dimensional representations that are easier to process for tasks such as recognition, tracking, and control. For flexible manipulators, feature representations must be robust against environmental noise (e.g., illumination changes and occlusions) while retaining sufficient detail to capture the manipulator's own deformations.

Classical approaches rely on handcrafted features such as SIFT, SURF, and HOG, which provide rotation- and scale-invariant representations. However, deep learning has gradually replaced these methods with data-driven feature learning, where convolutional neural networks (CNNs) and attention mechanisms automatically extract hierarchical features. Low-level features capture edges and textures, while high-level features encode semantic attributes, such as object categories and material properties.

For flexible manipulators, the challenge lies in extracting features that jointly represent object state and manipulator deformation. Multimodal feature fusion methods that integrate RGB, depth, and tactile cues have proven effective in building richer representations. Moreover, graph-based representations can model the structural relationships between manipulator joints and object surfaces, supporting more precise control strategies [31–33].

3.6 Visual Feedback Control

Visual feedback control integrates perception and control, allowing flexible manipulators to adjust their actions in real-time based on visual observations. Unlike rigid manipulators, flexible manipulators exhibit nonlinear and uncertain dynamics due to deformation, which makes visual feedback even more critical.

Two main paradigms exist: Position-based visual servoing (PBVS) and image-based visual servoing (IBVS). PBVS reconstructs the 3D pose of the target and aligns the manipulator accordingly, while IBVS directly minimizes image-plane errors without full 3D reconstruction. IBVS is generally more robust to calibration errors, whereas PBVS provides better global accuracy.

For flexible manipulators, hybrid approaches are often adopted. For example, PBVS provides coarse alignment in 3D space, while IBVS ensures fine adjustments at the image level. Advanced controllers further integrate adaptive and robust control techniques to compensate for nonlinearities, ensuring stability under uncertainties [34–36].

3.7 Visual-Tactile Fusion

Vision alone may not provide sufficient information in tasks involving deformable or textureless objects. Tactile sensing complements vision by directly measuring contact forces, slippage, and surface properties. Visual-tactile fusion allows flexible manipulators to achieve more natural and reliable interaction.

Fusion strategies can be categorized into early fusion (combining raw sensory data), intermediate fusion (integrating feature-level representations), and late fusion (decision-level integration). Recent work emphasizes deep multimodal learning, where neural networks jointly learn from visual and tactile modalities to produce unified representations.

For example, when grasping a soft fruit, vision can estimate shape and size, while tactile sensors measure deformation and firmness. Fusion of these cues enables the manipulator to adjust its grip force adaptively, avoiding both slippage and damage. Thus, visual-tactile fusion plays a central role in enabling safe, precise, and dexterous manipulation [37, 38].

3.8 Multi-camera Systems and Distributed Vision

Single-camera systems often suffer from occlusion and limited field of view, particularly in cluttered or collaborative environments. Multi-camera systems expand spatial coverage and enhance perception accuracy by providing redundant viewpoints.

In distributed vision setups, multiple cameras are strategically placed around the workspace, and their data are fused into a global representation. Key challenges include camera calibration, synchronization, and real-time data fusion. Recent methods employ simultaneous localization and mapping (SLAM) algorithms and deep multi-view fusion networks to address these issues.

For flexible manipulators, distributed vision is especially valuable for monitoring self-deformation. By observing the manipulator from multiple angles, the system can reconstruct its entire shape in real-time, improving accuracy in deformation compensation and collision avoidance [39, 40].

3.9 Visual Perception in Dynamic Environments

Dynamic environments introduce moving objects, varying illumination, and unpredictable human activities, all of which challenge visual perception. Flexible manipulators must not only perceive static scenes but also adapt to continuous changes.

Dynamic visual perception relies on algorithms capable of real-time tracking, background modeling, and motion segmentation. Deep recurrent networks, temporal attention mechanisms, and event-based cameras are increasingly used to handle rapid scene variations.

In human–robot collaboration scenarios, dynamic perception enables the manipulator to predict human motion trajectories and adjust its behavior accordingly. For example, it can proactively avoid collisions or assist human partners in cooperative assembly tasks. Such adaptability is key for safe and efficient deployment in unstructured environments [41, 42].

3.10 Visual SLAM and Environment Mapping

Simultaneous Localization and Mapping (SLAM) is critical for flexible manipulators operating in unknown or large-scale environments. By integrating visual input with odometry and inertial measurements, SLAM systems can build consistent maps while tracking manipulator pose.

Visual SLAM methods can be categorized into feature-based (using point or line features) and direct methods (using pixel intensity directly). Recent advances in deep learning-based SLAM further improve robustness under challenging conditions such as low texture or dynamic scenes.

For flexible manipulators, SLAM not only supports navigation but also facilitates manipulation planning. Accurate environment maps allow the manipulator to reason about obstacles, reachable regions, and optimal grasping positions. Moreover, deformable SLAM models that incorporate manipulator flexibility are emerging as a new research direction [43, 44].

3.11 Vision-Based Deformation Sensing

Traditional sensors such as strain gauges and encoders are difficult to deploy across the entire surface of a flexible manipulator. Vision-based deformation sensing provides a non-contact alternative for capturing global deformation states.

Techniques include marker-based tracking, where fiducial markers are attached along the manipulator body, and markerless methods, which rely on deep learning to estimate deformation directly from images. Depth cameras and structured light systems further enhance accuracy by capturing 3D surface geometries.

Vision-based deformation sensing is crucial for feedback control, as it provides real-time information about the manipulator's shape and posture. This enables closed-loop compensation strategies that maintain precision despite structural softness [45, 46].

3.12 Visual Learning and Self-Adaptation

Flexible manipulators must adapt to a wide variety of tasks and environments. Visual learning provides the foundation for such adaptability. By leveraging large datasets and self-supervised learning strategies, visual systems can continuously improve recognition, tracking, and control performance.

Key approaches include imitation learning, where the manipulator learns from human demonstrations, and reinforcement learning, where visual feedback guides policy optimization. Self-supervised learning further reduces dependence on labeled data by exploiting inherent structures in visual input.

Through visual learning, flexible manipulators acquire the ability to generalize across unseen scenarios. For instance, after training on grasping cylindrical objects, they may successfully transfer strategies to grasp bottles or cans with similar features. This adaptability greatly enhances autonomy in real-world applications [47–49].

3.13 Challenges and Future Directions

Despite significant progress, several challenges remain in visual perception for flexible manipulators. Real-time performance: High-resolution visual processing and deep learning models demand considerable computational resources. Robustness in complex environments: Illumination variations and occlusions still degrade perception accuracy. Deformation-aware perception: Integrating deformation models remains an open problem. Scalability and generalization: Learning-based systems struggle with domain transfer. Human–robot collaboration: Dynamic perception and intent recognition remain active research areas. Future research is expected to focus on multimodal integration, bio-inspired perception mechanisms, and neuro-symbolic learning frameworks [50, 51].

References

1. Wu, J., & Zhou, L. (2023). Transcending the acceleration-bandwidth trade-off: Lightweight precision stages with active control of flexible dynamics [Preprint]. *arXiv*. INSPEC: 23783913.
2. Okunevich, I., Trinitatova, D., Kopanev, P., & Tsetserukou, D. (2021). MobileCharger: An autonomous mobile robot with inverted delta actuator for robust and safe robot charging. In *2021 26th IEEE International Conference on Emerging Technologies and Factory Automation (ETFA)*. IEEE. WOS: 000766992600090. https://doi.org/10.1109/ETFA45728.2021.9613366
3. Arumugom, S., Muthuraman, S., & Ponselvan, V. (2009). Modeling and application of series elastic actuators for force control multi legged robots. *Journal of Computing, 1*(1) ISSN: 2151-9617.

4. Liu, G., Kurnia, A., De Larminat, R., & Rotter, S. J. (2003). Position sensor error analysis for EPS motor drive. In *IEEE International Electric Machines and Drives Conference, 2003. IEMDC'03* (Vol. 1, pp. 249–254). IEEE. https://doi.org/10.1109/IEMDC.2003.1211271
5. Farronato, M., Mannocci, P., Milozzi, A., Compagnoni, C. M., & Ielmini, D. (2025). Bio-inspired computing with emerging devices: Bridging 2D materials and neuromorphic architectures. In *2025 IEEE Computer Society Annual Symposium on VLSI (ISVLSI)* (pp. 1–1). IEEE. https://doi.org/10.1109/ISVLSI65124.2025.11130347
6. Ju, Q., & Pan, F. (2016). Performance comprehensive assessment system for the manufacturing system of automotive assembly line. In *2016 Chinese Control and Decision Conference (CCDC)* (pp. 4672–4676). IEEE. https://doi.org/10.1109/CCDC.2016.7531827
7. Vasilev, V. E. (2025). Examining the productivity of agent trained with RL and IL for emergency tasks in a virtual electrical substation. In *2025 60th International Scientific Conference on Information, Communication and Energy Systems and Technologies (ICEST)* (pp. 1–4). IEEE.
8. Wu, R., Qin, C., Huang, G., Tao, J., & Liu, C. (2023). Long term prediction of hydraulic oil properties with sample convolution and interaction networks. In *2023 9th international conference on Fluid Power and Mechatronics (FPM)* (pp. 1–5). IEEE. https://doi.org/10.1109/FPM57590.2023.10565519
9. Esfandiari, M., Gehlbach, P., Taylor, R. H., & Iordachita, I. I. (n.d.). Bimanual manipulation of steady-hand eye robots with adaptive sclera force control: Cooperative versus teleoperation strategies. *IEEE Transactions on Human-Machine Systems*. https://doi.org/10.1109/THMS.2025.3605011
10. Kim, D. H., Heo, S.-H., & Park, H.-S. (2017). Biomimetic finger extension mechanism for soft wearable hand rehabilitation devices. In *2017 International Conference on Rehabilitation Robotics (ICORR)* (pp. 1326–1330). IEEE. https://doi.org/10.1109/ICORR.2017.8009432
11. Glebov, N., Kruglova, T., & Shoshiashvili, M. (2019). Intelligent electro-pneumatic module for industrial robots. In *2019 international multi-conference on industrial engineering and modern technologies (FarEastCon)* (pp. 01–04). IEEE. https://doi.org/10.1109/FarEastCon.2019.8934864
12. Diao, W., Pan, Q., Li, Y., Zhang, J., Liu, P., Chen, J., & Feng, Z. (2022). Dynamic investigation on effective bulk modulus of hydraulic oil using piezoelectric transducers. In *2022 16th Symposium on Piezoelectricity, Acoustic Waves, and Device Applications (SPAWDA)* (pp. 614–618). IEEE. https://doi.org/10.1109/SPAWDA56268.2022.10046001
13. Zhao, F., Wen, Y. K., Wen, X. H., Zhao, L., & Guo, Q. J. (2014). Optimal design of planetary gear in multi-mode hybrid drive system. In *2014 IEEE conference and expo transportation electrification Asia-Pacific (ITEC-Asia-Pacific)* (pp. 1–5). IEEE. https://doi.org/10.1109/ITEC-AP.2014.6940629
14. Zhang, L., Shen, Z., Peng, Q., Yang, L., & Bao, Y. (2020). A novel low-energy input welding system for lightweight material. In *2020 7th International Conference on Information Science and Control Engineering (ICISCE)* (pp. 1585–1589). IEEE. https://doi.org/10.1109/ICISCE50968.2020.00313
15. Flores, C. K. F. (2021). Design, modeling, and simulation of a wheeled, wall-pressed, in-pipe inspection robot for pipes with 6-8 inches inside diameter. In *2021 IEEE International Conference on Automatic Control & Intelligent Systems (I2CACIS)* (pp. 235–240). IEEE. https://doi.org/10.1109/I2CACIS52118.2021.9495903
16. Els, D., McLaren, J., van Niekerk, T., & Phillips, R. (2023). Design, fabrication and analysis of a soft robotic gripper using fluid elastomer actuators. In *2023 14th International Conference on Mechanical and Intelligent Manufacturing Technologies (ICMIMT)* (pp. 1–7). IEEE. https://doi.org/10.1109/ICMIMT59138.2023.10199541
17. Mielke, T., Göpfert, J.-M. W., & Murrenhoff, H. (2018). Potential of ferromagnetic shape memory alloys as power units in hydraulic pumps. In *2018 Global Fluid Power Society PhD symposium (GFPS)* (pp. 1–7). IEEE. https://doi.org/10.1109/GFPS.2018.8472391
18. Zhang, Y., Shi, J., Wang, S., Zhang, C., & Tomovic, M. M. (2016). Sealing mechanism and failure analysis of actuator reciprocating seal. In *2016 IEEE 11th Conference on Industrial*

Electronics and Applications (ICIEA) (pp. 2190–2195). IEEE. https://doi.org/10.1109/ICIEA.2016.7603952

19. Fenjan, S. Q., & Dehkordi, S. F. (2023). Design and fabrication of a pneumatic soft robot gripper using hyper-flexible silicone. In *2023 11th RSI International Conference on Robotics and Mechatronics (ICRoM)* (pp. 641–646). IEEE. https://doi.org/10.1109/ICRoM60803.2023.10412446
20. O'Neill, P. E., Lang, R. H., Kurum, M., Utku, C., & Carver, K. R. (2006). Multi-sensor microwave soil moisture remote sensing: NASA's combined radar/radiometer (ComRAD) system. In *2006 IEEE MicroRad* (pp. 50–54). IEEE. https://doi.org/10.1109/MICRAD.2006.1677061
21. Chen, J., Li, Y., & Zhang, H. (2021). Vision-based adaptive control of soft manipulators. *IEEE Transactions on Robotics*.
22. Renda, F., et al. (2018). Discrete Cosserat approach for multisection soft manipulator dynamics. *IEEE Transactions on Robotics, 34*, 1518.
23. Eigen, D., Puhrsch, C., & Fergus, R. (2014). *Depth map prediction from a single image using a multi-scale deep network*. NeurIPS.
24. Ranftl, R., Lasinger, K., Hafner, D., et al. (2021). *Vision transformers for dense prediction*. IEEE ICCV.
25. Yin, X., et al. (2023). Monocular depth estimation via multi-scale fusion. *IEEE TMM*.
26. Dosovitskiy, A., Beyer, L., Kolesnikov, A., et al. (2021). *An image is worth 16x16 words: Transformers for image recognition at scale*. ICLR.
27. He, K., Zhang, X., Ren, S., & Sun, J. (2016). *Deep residual learning for image recognition*. IEEE CVPR.
28. Radford, A., Kim, J. W., Hallacy, C., et al. (2021). *Learning transferable visual models from natural language supervision*. ICML.
29. Bewley, A., Ge, Z., Ott, L., Ramos, F., & Upcroft, B. (2016). *Simple online and realtime tracking*. IEEE ICIP.
30. Danelljan, M., Gool, L. V., & Timofte, R. (2020). *Probabilistic regression for visual tracking*. IEEE CVPR.
31. Lowe, D. (2004). Distinctive image features from scale-invariant keypoints. *International Journal of Computer Vision, 60*, 91.
32. Dalal, N., & Triggs, B. (2005). *Histograms of oriented gradients for human detection*. IEEE CVPR.
33. Vaswani, A., Shazeer, N., Parmar, N., et al. (2017). Attention is all you need. *NeurIPS*.
34. Hutchinson, S., Hager, G. D., & Corke, P. (1996). A tutorial on visual servo control. *IEEE Transactions on Robotics and Automation, 12*, 651.
35. Chaumette, F., & Hutchinson, S. (2006). Visual servo control. *IEEE Robotics & Automation Magazine*.
36. Kragic, D., & Christensen, H. (2002). *Survey on visual servoing for manipulation*. CVAP Report.
37. Calandra, R., Owens, A., Jayaraman, D., Lin, J., Yuan, W., Adelson, E. H., & Levine, S. (2018). More than a feeling: Learning to grasp and regrasp using vision and touch. *IEEE Robotics and Automation Letters, 3*, 3300.
38. Lee, M. H., et al. (2019). Vision-based tactile sensing. *Sensors, 19*(7), 1612.
39. Newcombe, R. A., Izadi, S., et al. (2011). *KinectFusion: Real-time dense surface mapping and tracking*. ISMAR.
40. Li, Y., Wang, F., & Zhang, C. (2020). Distributed vision system for multi-robot collaboration. *IEEE Access*.
41. Zhu, A., Yuan, L., Chaney, K., & Daniilidis, K. (2017). *Unsupervised event-based learning of optical flow*. IEEE CVPR.
42. Gallego, G., Delbruck, T., Orchard, G. M., et al. (2020). Event-based vision: A survey. *IEEE TPAMI*.
43. Mur-Artal, R., & Tardos, J. D. (2017). ORB-SLAM2: An open-source SLAM system for monocular, stereo, and RGB-D cameras. *IEEE Transactions on Robotics, 33*, 1255.

44. Bloesch, M., Czarnowski, J., Clark, R., Leutenegger, S., & Davison, A. J. (2018). *CodeSLAM—Learning a compact representation for dense visual SLAM*. IEEE CVPR.
45. Thuruthel, T. G., Shih, B., Laschi, C., & Tolley, M. T. (2019). Soft robot perception using deep learning. *Soft Robotics, 6*(3), 399–413.
46. Wang, H., Liu, J., & Zhu, Q. (2022). Vision-based shape estimation for continuum robots. *Robotics and Computer-Integrated Manufacturing, 74*, 102289.
47. Levine, S., Finn, C., Darrell, T., & Abbeel, P. (2016). End-to-end training of deep visuomotor policies. *Journal of Machine Learning Research*.
48. Finn, C., Levine, S., & Abbeel, P. (2017). *Deep visual foresight for planning robot motion*. IEEE ICRA.
49. Zhan, A., Lee, Y., & Wang, Z. (2021). *Self-supervised visual learning for robotic manipulation*. IEEE ICRA.
50. Kim, S., Laschi, C., & Trimmer, B. (2016). Soft robotics: A bioinspired evolution in robotics. *Science, 354*(6315), 1248–1252.
51. Rus, D., & Tolley, M. (2015). Design, fabrication and control of soft robots. *Nature, 521*(7553), 467–475.

Control Methods for Flexible Manipulators

1 Control Theory and Methods

Optimal compromise between rigid-body motion tracking and elastic vibration suppression through careful design of the state weighting matrices, providing guaranteed stability and a degree of robustness. However, the principal limitation of classical control methods lies in their heavy reliance on an accurate mathematical model of the plant; the performance of controllers based on linearized models degrades significantly during large-range maneuvers or in the presence of parameter uncertainties and external disturbances.

To overcome the limitations of classical control, model-based advanced control strategies emerged, among which computed torque control (also known as inverse dynamics control) and variable structure control are highly representative. The core idea of computed torque control is to utilize the system's nonlinear dynamic model for feedforward compensation, effectively canceling out the nonlinear coupling terms. Specifically, the controller computes in real-time the torque required to produce the desired acceleration, which encompasses all nonlinear terms, including inertial, Coriolis, centrifugal, and gravitational forces. This is then combined with a simple linear feedback law (e.g., PD control) to stabilize the system and eliminate tracking errors. Ideally, this method decouples and linearizes the complex nonlinear system into a set of simple double-integrator systems, enabling high-precision tracking. However, its success is critically dependent on the accuracy of the dynamic model. For flexible manipulators, obtaining a perfectly accurate inverse dynamics model is extremely challenging, and any model uncertainties directly translate into tracking errors.

Variable structure control, particularly sliding mode control (SMC), is renowned for its strong robustness to parameter uncertainties and external disturbances. The design of SMC involves two steps: first, designing an ideal sliding surface such that once the system state trajectories reach this surface, their dynamics will converge to

J. Qu, *Robotic Flexible Manipulation and Intelligent Sensing Technologies*,
https://doi.org/10.1007/978-981-95-6232-9_4

the equilibrium point with predetermined performance; second, designing a discontinuous control law that forces the system states to reach and remain on the sliding surface within finite time. The robustness of this strategy is evidenced by its ability to guarantee system stability provided the uncertainties are matched and bounded in norm. For flexible manipulators, SMC can effectively handle variations in model parameters, such as payload mass and joint friction. Nevertheless, the inherent "chattering" phenomenon—high-frequency switching of the control signal—is its major drawback, which may excite unmodeled high-frequency dynamics and cause actuator wear. Numerous methods, such as the boundary layer technique and higher-order sliding modes, have been developed to attenuate chattering.

With increasing demands on control performance, intelligent control methods and adaptive/robust control methods have demonstrated significant potential. Intelligent control, encompassing fuzzy logic control, neural network control, and their hybrids, does not rely on precise mathematical models but instead utilizes human expertise or data-driven approaches to handle system nonlinearities and uncertainties. Fuzzy logic controllers encapsulate expert control knowledge into "IF-THEN" rules, enabling effective control of complex objects like flexible manipulators, especially when accurate models are difficult to derive. Neural networks, leveraging their powerful nonlinear approximation and learning capabilities, can be employed to identify the system's inverse dynamics online or offline, forming neuro-adaptive controllers, or can act directly as controllers. While these methods excel in handling high nonlinearities and uncertainties, their theoretical analysis, such as stability proof, is often challenging, and their design relies heavily on substantial data and expert knowledge.

Adaptive control and robust control address uncertainties from a systems theory perspective. Adaptive control automatically adjusts controller parameters in real-time to accommodate slow variations in system dynamics, such as payload changes. Model Reference Adaptive Control (MRAC) is a typical approach, where an adaptation law is designed so that the actual output of the flexible manipulator asymptotically tracks the output of a reference model with desired performance characteristics. Robust control, exemplified by H-infinity control, aims not to make the system insensitive to uncertainties, but to minimize the worst-case effect of uncertainties on system performance, quantified by the H-infinity norm of the transfer function from disturbances to outputs. This implies that worst-case uncertainties are considered during the design phase, ensuring stability and satisfactory performance under all anticipated uncertainties. H-infinity control is particularly suited for complex environments with simultaneous model uncertainties, external disturbances, and measurement noise, although the resulting controller order is often high, complicating implementation.

Looking forward, control theories for flexible manipulators are evolving toward greater integration, intelligence, and performance orientation. Hybrid control strategies are becoming a trend—for instance, combining adaptive fuzzy control with sliding mode control to retain robustness while using fuzzy systems to approximate uncertainties and reduce chattering, or integrating neural networks within an H-infinity framework to ensure robust stability in a data-driven manner. Furthermore, advancements in sensor technology facilitate multisensor data fusion, combining information from motor encoders, end-effector vision systems, and strain gauges or accelerometers on the link, which enables more comprehensive state observation and precise control. Observer-based control methodologies, such as using an Extended State Observer (ESO) to estimate and compensate in real-time for total disturbances encompassing both internal dynamics and external disturbances, as demonstrated by Active Disturbance Rejection Control (ADRC), offer a new pathway that reduces dependence on precise models.

In summary, the control theories and methods for flexible manipulators represent a continuous evolutionary process, progressing from simple to complex, linear to nonlinear, and from model-dependent to uncertainty-tolerant and even learning-based approaches. No single method is universally optimal; the selection for practical applications invariably involves a trade-off among control performance, computational complexity, robustness, and implementation cost. A deep understanding of the core principles, advantages, and limitations of each methodology is crucial for designing and implementing efficient and reliable control systems tailored to specific application scenarios.

2 Intelligent Control Algorithms

In the field of flexible manipulator control, intelligent control algorithms have been widely studied and applied in recent years as effective strategies to address nonlinearity, high dimensionality, strong coupling, and uncertainty. These methods are mainly categorized into fuzzy control, neural network control, intelligent optimization algorithm control, and reinforcement learning control.

2.1 Fuzzy Logic Control

Fuzzy Logic Control (FLC) is a control technique based on fuzzy set theory, linguistic variables, and inference rules. Instead of relying on an accurate mathematical model, it uses expert-defined IF-THEN rules to handle complex, uncertain, and nonlinear systems. In flexible manipulators, FLC is especially useful for dealing with dynamic uncertainties and strong nonlinearities caused by joint flexibility and external disturbances. By fuzzifying input variables such as position error and its derivative, and applying fuzzy inference to determine control actions, the method provides a robust control framework suitable for real-time trajectory tracking and disturbance rejection.

This study designs a fuzzy logic controller (FLC) for the trajectory control of a robot manipulator equipped with a rotating-prismatic (RP) joint and a sliding flexible arm. The primary objective of this research is to achieve precise positioning of the end-effector tip while simultaneously suppressing the residual vibrations of the flexible arm that arise during motion. The manipulator is required to follow a multi-segment straight-line trajectory, which presents challenges due to the nonlinear dynamics and elastic deformation inherent in flexible-link manipulators.

Simulation results (and potentially experimental validation) demonstrate that the fuzzy logic controller significantly improves tracking performance and reduces oscillations compared to conventional PID control or model-based techniques. This approach provides a flexible and adaptive framework for real-time control of flexible robotic manipulators in applications such as precision assembly, space robotics, and micro-manipulation tasks [1].

2.2 *Neural Network Control*

Neural Network Control (NNC) leverages the nonlinear function approximation capabilities of multilayer neural networks to learn the input-output relationships of a system from data. It is especially effective for systems with complex dynamics, unknown models, or strong parameter coupling. In flexible manipulators, NNC can be applied for system identification, inverse dynamics modeling, and adaptive trajectory tracking. By learning the system's behavior through training, the neural controller can compensate for nonlinearities, suppress vibrations, and adapt to external disturbances. Common architectures include BP networks, radial basis function (RBF) networks, and convolutional neural networks (CNNs).

This study employs adaptive neural networks to design controllers for suppressing vibrations of a flexible manipulator, significantly improving the accuracy in describing its elastic deflection and accomplishing complex closed-loop control tasks [2].

2.3 *Fuzzy Neural Network Control*

Fuzzy neural network (FNN) is a hybrid intelligent control strategy that combines the reasoning framework of fuzzy logic with the learning ability of neural networks. It enables adaptive tuning of fuzzy rules, membership functions, and control parameters through a data-driven training process. In flexible manipulators, FNNs are widely used for vibration suppression, joint flexibility compensation, and adaptive trajectory tracking. By integrating the robustness of fuzzy logic and the adaptability of neural networks, FNNs offer enhanced performance in controlling highly nonlinear and uncertain systems.

This study proposes a robust adaptive fuzzy neural controller (AFNC) for motion control of multilink soft robotic manipulators, enabling rapid tracking of desired trajectories by the robot joints while maintaining high control performance in the presence of external disturbances [3].

2.4 Intelligent Optimization Algorithm Control

Intelligent optimization algorithm control refers to a class of global search optimization techniques inspired by natural phenomena, such as evolution, swarm intelligence, or physical processes. These algorithms include, but are not limited to, genetic algorithms (GA), particle swarm optimization (PSO), ant colony optimization (ACO), differential evolution (DE), and simulated annealing (SA). In the context of flexible manipulator control, intelligent optimization algorithms are widely used for tuning fuzzy rule parameters, optimizing the weights and architecture of neural networks, adapting controller gains, and designing multi-objective control laws. Compared with traditional gradient-based methods, these algorithms exhibit strong global optimization capabilities and robustness, making them particularly suitable for systems that are nonlinear, high-dimensional, constrained, or non-differentiable. Although some algorithms may involve high computational costs or slower convergence, their practical performance can be greatly improved through the incorporation of heuristic rules, hybrid local search mechanisms, or parallel computing frameworks. Recently, an increasing number of studies have integrated intelligent optimization algorithms with fuzzy systems, neural networks, and reinforcement learning to develop high-performance hybrid control strategies for flexible manipulators.

This study addresses soft robotic manipulators in industrial environments by using intelligent optimization algorithms to solve inverse kinematics, enabling the manipulator to flexibly avoid obstacles and successfully navigate through multi-obstacle environments [4].

2.5 Reinforcement Learning Control

Reinforcement learning (RL) is a reward-driven sequential decision-making method where an agent learns optimal control strategies through continuous interaction with the environment. It does not rely on an accurate model of the system, making it suitable for high-dimensional, continuous control problems. In flexible manipulators, RL can be used for learning motion strategies, optimizing dynamic trajectories, and improving end-effector accuracy. Recent advances in deep reinforcement learning (e.g., DDPG, TD3, and SAC) integrate neural networks for function approximation, enabling effective control in complex, multi-joint, and uncertain environments.

This study employs a neural network-based modeling approach to construct an approximate dynamic model of a cable-driven soft manipulator, a system characterized by highly nonlinear, coupled, and time-varying dynamics. Due to the manipulator's inherent structural flexibility and the complex interaction between the cables and the soft body, obtaining an accurate analytical model is extremely challenging. To overcome this limitation, a data-driven neural network model is developed to learn the nonlinear mapping between the actuator inputs, internal states, and the resulting end-effector motion.

The proposed framework demonstrates strong adaptability to unmodeled dynamics and external disturbances, highlighting its potential for real-time, high-performance control of soft robotic systems in applications such as minimally invasive surgery, human-robot interaction, and flexible manipulation tasks [5].

2.6 Expert Systems and Hybrid Intelligent Control

Expert systems are intelligent control frameworks that mimic human decision-making by utilizing rule-based reasoning and knowledge bases. In flexible manipulators, they are often employed for fault-tolerant control, task scheduling, or system diagnosis. To enhance adaptability and control performance, expert systems are frequently combined with fuzzy logic, neural networks, genetic algorithms, and other techniques to form hybrid intelligent controllers. These systems integrate symbolic reasoning and data-driven learning, enabling effective control in complex, dynamic environments with minimal reliance on precise system models.

This study proposes a hybrid intelligent control strategy that integrates neural networks, fuzzy logic, and particle swarm optimization (PSO) to achieve high-precision control of a two-degree-of-freedom (2-DOF) robotic arm within a flexible robot system.

The integration of these three components enables the robotic arm to achieve smooth motion, rapid convergence, and precise positioning of the claw, while effectively compensating for nonlinearities and coupling effects between the joints. Simulation and experimental results demonstrate that the proposed hybrid controller outperforms conventional PID and standalone intelligent control methods in terms of trajectory tracking accuracy, stability, and vibration suppression.

This hybrid control framework provides a promising solution for complex robotic systems that require high flexibility and precision, with potential applications in industrial automation, medical robotics, and delicate manipulation tasks [6].

3 Multimodal Fusion Control

3.1 Introduction: Reasons for Using Multimodal Fusion Control in Flexible Arms

In the preceding discussion, we explored the applications and challenges of traditional control algorithms and intelligent control algorithms in the field of flexible arm control. Single-modal control algorithms, whether traditional or intelligent, have significant limitations. Traditional control theory primarily relies on model-based control methods, which are fundamentally based on modeling the flexible arm using a single-modal approach, such as the Cosserat rod theory for flexible arm modeling. Single-mode intelligent algorithms, such as reinforcement learning and deep learning, exhibit certain advantages over traditional control algorithms in actual control performance. However, the training data for intelligent algorithms is constrained by single-mode data sources, making it difficult to guarantee their control effectiveness. Compared to traditional rigid robotic arms, flexible robotic arms have structural characteristics such as infinite degrees of freedom, uncontrollable elastic deformation, and complex force coupling. When performing high-precision operational tasks, they require extremely high levels of perception capabilities and control strategies.

In current research in this field, most flexible arm control algorithms typically rely on single-modal sensory information, such as image-based target recognition and path planning, or force sensors for contact detection. The actual application scenarios for flexible arms are generally characterized by unstructured environments, strong dynamic characteristics, and frequent interference. Single-modal control struggles to provide stable and comprehensive environmental perception or feedback on the flexible arm's own state. For example, image data is easily affected by environmental lighting and obstructions, while force sensors are often single-point or small-area sensors, making it difficult to provide global information. Single-modal control suffers from incomplete information, which significantly impacts the accuracy of control algorithms and system robustness. This drawback is particularly pronounced in complex environments and those with interactive characteristics.

To overcome the aforementioned shortcomings, multimodal control has emerged. In recent years, multimodal fusion control has gradually become a popular research direction in the field of flexible arm control algorithms. Multimodal fusion control refers to the integration of data from multiple sensors, such as visual, force, tactile, posture, and audio sensors, to fuse information from multiple sensory sources. By utilizing data fusion algorithms for unified modeling and decision-making control, this approach enhances the intelligence, adaptability, and robustness of flexible arm control. This, in turn, improves the performance of flexible arms in complex environments and those with interactive characteristics, enabling precise operations and complex interactions.

This section will focus on the basic concepts of multimodal fusion control, perception mode analysis, information fusion algorithms, and control architecture and will explore its typical applications in the field of flexible arm control as well as the challenges currently facing technological development.

Given the structural characteristics of flexible arms, the perception system must have the ability to obtain real-time multidimensional information on the environmental state, target object attributes, and the state of the robotic arm itself in order to enable the flexible arm to achieve high-precision operation and stable control in complex tasks. Unlike traditional rigid robotic arms, which primarily rely on position sensors or joint encoders to detect posture, flexible arms have significant nonlinear deformation characteristics and infinite degrees of freedom, thereby requiring a higher demand for perception dimensions. This has led to the emergence of a "multimodal perception" architecture, which involves integrating multiple sensor modalities to work in tandem, providing the control system with comprehensive and robust perception data.

In the multimodal perception system of a flexible arm, the visual modality plays a crucial role. Through RGB cameras, depth cameras, or stereo vision systems, the robot can obtain geometric structure information of the task environment, as well as the position and orientation of target objects. This non-contact perception method is suitable for long-distance perception and path planning, but its stability is easily affected by factors such as occlusion and changes in lighting conditions. Therefore, the visual modality typically needs to be used in conjunction with other modalities to enhance system robustness.

The force modality provides contact feedback between the flexible arm and objects/the environment, making it an indispensable component of flexible arm control. By installing six-degree-of-freedom force or torque sensors on the end-effector or joints, the system can monitor the forces and torques applied during grasping in real-time, thereby adjusting grasping strategies to avoid damaging objects. Compared to vision, force modality more directly reflects the "interaction state," particularly in tasks such as dynamic grasping and assembly.

Tactile modality, however, focuses more on detailed information at the contact point, such as contact area, pressure distribution, friction force, and surface texture. Represented by capacitive, piezoresistive, or fiber-optic flexible tactile sensors, tactile modality is primarily used in actuator components such as fingertips or flexible grippers, enhancing the precision and safety of grasping. Especially in tasks involving contact with human skin, handling soft objects, or micro-manipulation, tactile information can provide critical perceptual data that traditional visual and force modalities struggle to obtain.

The inertial modality uses an IMU (inertial measurement unit) to obtain data, such as acceleration and angular velocity, which can be used to estimate the robot arm's posture, motion trends, and structural vibration status. For flexible arms, IMUs can be deployed on multiple joints or flexible-link structures to help the system perceive nonlinear deformation processes, construct dynamic models, and is particularly suitable for real-time posture control and vibration detection.

Additionally, in semantic tasks such as service robots and human-robot collaboration, voice modalities may be introduced as auxiliary inputs for parsing human commands, intent recognition, or interactive instructions. Although the voice modality does not directly participate in control variable acquisition, it plays a significant role in strategy selection or state switching. The ability to recognize semantic information enhances the system's interactivity and flexibility.

Sensor perception data from different modalities each have their own characteristics: vision provides global spatial observation information, force and touch provide physical interaction feedback, IMU focuses on the arm's own motion state, and voice is suitable for high-level interaction command analysis. They each have advantages and disadvantages in terms of spatiotemporal resolution, real-time performance, and robustness. Therefore, by reasonably combining and fusing these modalities, the perception capabilities of the flexible robotic arm control system can be "complementarily optimized." The next subsection will further discuss how these modal information can be fused at multiple levels to construct a more intelligent and efficient control framework.

3.2 Multimodal Data Fusion Mechanism

The core issue in multimodal fusion control lies in how to integrate sensor information from diverse sources with distinct characteristics into a unified control input, thereby achieving efficient and stable control decisions. In the control system of a flexible arm, sensors of different modalities (such as vision, force, touch, IMU, etc.) typically output data at different sampling frequencies, timestamps, precisions, and dimensions, which imposes higher requirements on the system's perception consistency and decision coherence. To fully leverage the synergistic effects of multi-source perception, it is essential to establish a scientific and systematic data fusion mechanism, with careful design required across processing layers, spatiotemporal alignment strategies, and fusion method selection.

Multimodal data fusion is typically categorized into three types based on processing layers: data-layer fusion, feature-layer fusion, and decision-layer fusion. Data-layer fusion is the most basic approach, directly concatenating or overlaying raw sensor data, requiring high temporal and spatial synchronization among multimodal data. This method offers fast fusion speed and strong real-time performance, commonly used for visual-IMU joint localization (e.g., VIO), but it demands extremely high synchronization accuracy and is sensitive to noise. Feature-level fusion involves extracting key features from preprocessed data across modalities, then using neural networks, attention mechanisms, or graph structures for fusion modeling. This approach balances efficiency and expressive capability, making it suitable for complex control tasks. Decision-level fusion processes data from each modality independently to make preliminary judgments, then merges them at the strategy level. This is the loosest-structured and most fault-tolerant fusion method, suitable for scenarios with high task uncertainty or significant delays between modalities.

In addition to fusion layer design, spatiotemporal synchronization is the most challenging aspect of multimodal fusion in engineering. Different sensors have significantly different sampling frequencies, such as visual cameras typically operating at 60 Hz, IMUs commonly at 200 Hz, and force sensors as high as 1000 Hz. Additionally, different modalities often operate in distinct spatial coordinate systems, such as image space, global coordinate systems, or end-effector reference frames of robotic arms. These asynchronies result in misalignment of data in both the temporal and spatial representations. To address this, methods such as timestamp alignment, interpolation resampling, and multi-frequency fusion are employed to unify the data. Spatial calibration relies on sensor calibration using external parameters and coordinate transformation matrices to ensure that data from all modalities are expressed in a unified spatial representation of the same task state.

Different fusion methods are suitable for different task requirements and system constraints. Generally, high-real-time tasks such as dynamic grasping or obstacle avoidance are suitable for data-layer fusion; while tasks with higher requirements for control accuracy and semantic understanding, such as precision assembly or human-machine interaction, are more suitable for feature-layer or decision-layer fusion. Additionally, the selection of fusion methods must also consider factors such as modal synchronicity, system computational resources, and delay tolerance of the control loop. In engineering practice, it is often necessary to balance robustness, real-time performance, and computational complexity to design a fusion framework that is both practical and scalable.

In recent years, with the development of deep learning and graph modeling techniques, multimodal fusion mechanisms are transitioning from static structures to dynamic, adaptive modeling. For example, multimodal transformers dynamically assign weights to the importance of different modalities through self-attention mechanisms; graph neural networks explicitly model semantic and structural relationships between modalities to enable cross-modal message propagation; and strategies such as multi-task learning and self-supervised learning provide stronger generalization and fault tolerance capabilities for modal representation learning. The introduction of these technologies opens new pathways for autonomous perception and control of flexible robotic arms in uncertain environments.

In summary, multimodal data fusion mechanisms form the technical foundation for constructing intelligent control systems for flexible robotic arms. They not only enhance the completeness and robustness of perception but also provide high-quality input for subsequent control strategies and motion planning. The next section will further explore how to convert fused multimodal perception information into control strategies and algorithms, achieving a closed-loop system transition from "perception collaboration" to "control intelligence."

The ultimate goal of multimodal fusion control is to effectively convert perception information from different sensors into control commands for a flexible robotic arm, thereby enabling precise execution of various tasks in complex environments. After data fusion, the control system must perform multilayer processing on the fused information, including strategy planning, trajectory prediction, and motion adjustment. This requires the control architecture to possess excellent modularity, responsiveness, and adaptability. Therefore, current multimodal fusion control systems generally adopt a three-layer hierarchical structure comprising perception, decision-making, and execution layers. This ensures clear system logic, scalable structure, and facilitates the integration of corresponding algorithm models into each layer.

In this hierarchical control framework, the perception layer integrates information from modalities such as vision, force, touch, and IMU to provide the system with comprehensive observational data about the environment, targets, and arm state; the decision layer uses this information for task judgment, action selection, and trajectory planning, serving as the core decision-making hub of the system; the execution layer ultimately converts high-level strategies into control commands for the end-effector or joints to execute specific operational tasks. This closed-loop structure of "perception-driven decision-making followed by precise execution" has become a typical framework for multimodal control.

In terms of control strategies, different tasks and environments impose varying demands on the system's control capabilities, necessitating the design of tailored algorithms based on fused information. Model predictive control (MPC) is a widely applied optimization-based control method whose core principle involves predicting future system changes based on the current state and planning an optimal control sequence. In multimodal control, MPC can utilize fused visual, force, and IMU state information to achieve more accurate path prediction and dynamic obstacle avoidance control. For example, in an assembly scenario, the system can use vision for target localization, force sensing for contact state detection, and IMU for motion posture estimation, thereby forming a multimodal-driven predictive model to enhance control accuracy and robustness.

In addition to explicit model control strategies, an important direction is dynamic modal weighting control based on modal reliability. Since different modalities exhibit varying stability and practicality during task execution—e.g., vision is susceptible to occlusion, and force sensing may be affected by noise—the system must possess the ability to automatically adjust modal weights based on the current environment. Such adaptive weighting mechanisms can be implemented using Bayesian estimation, modal confidence assessment, or attention mechanism networks, enabling dynamic regulation of multimodal inputs and enhancing the flexibility and stability of control strategies in complex environments.

In recent years, deep reinforcement learning (DRL) has gradually become a research hotspot in the field of fusion control. DRL uses multimodal perception results as system state inputs and leverages policy networks to learn action distributions, thereby achieving end-to-end control from perception to action. This method is particularly suitable for flexible robotic arms performing tasks in complex or unstructured environments, such as autonomous grasping, multi-object tracking, and human-robot collaboration. By introducing multi-branch neural network structures, modality fusion encoders, and reward mechanism optimization, DRL can efficiently capture temporal dependencies and semantic associations between modalities, enabling adaptive control in complex environments.

In addition to the control strategy itself, multimodal control systems must also address issues such as sensor failure or unreliable modalities. To this end, many systems have introduced modality failure detection and robust control mechanisms. For example, abnormal modalities can be identified through modal health diagnosis mechanisms, or modal degradation strategies can be designed, such as automatically switching to force-based or IMU-based control methods when visual modalities fail. Additionally, the fusion algorithms themselves can incorporate robust design, such as introducing minimum variance estimation or fuzzy control methods to mitigate the impact of abnormal data, ensuring the system remains stable even in the absence of modalities or inconsistent information.

When evaluating the effectiveness of fusion control strategies, it is typically necessary to assess the performance of the control system from multiple dimensions. These include control accuracy (e.g., trajectory error and positioning deviation), system real-time performance (delay from perception to execution), robustness (stable response capability to abnormal inputs), adaptability (ability to transfer to new environments or tasks), and computational efficiency. These metrics not only reflect the engineering feasibility of the control algorithm but also determine its deployment capability in practical applications.

In summary, multimodal fusion control strategies are a key component in achieving deep coupling between perception and control. Through the synergistic effects of perception-driven modeling, dynamic fusion at the strategy layer, deep reinforcement learning empowerment, and modal failure tolerance mechanisms, flexible robotic arms can demonstrate stronger adaptability and intelligence when faced with uncertain tasks and dynamic environments. The next section will further introduce the application performance and actual effects of these control strategies in specific tasks based on typical scenarios.

Multimodal fusion control technology has become a key enabler for flexible robotic arms to achieve intelligent operations, environmental adaptation, and safe collaboration. In practical applications, control strategies that integrate multimodal signals such as vision, force, touch, IMU, voice, and electromyography can endow the system with comprehensive observational capabilities for complex task states and enable flexible adjustments to actions and paths in dynamic environments, thereby significantly enhancing the system's adaptability and robustness. The following sections will introduce the specific applications and technical characteristics of multimodal fusion control in flexible arms based on several representative scenarios.

In object grasping and manipulation tasks involving flexible arm end-effectors, the flexible arm must address challenges such as varying target shapes and complex force contacts. In such tasks, visual modalities can be used to provide information on object pose, boundaries, and surfaces, while force sensors can provide feedback on contact forces and slippage trends during the grasping process. Based on this, researchers have developed a control framework that integrates visual and force-sensing modalities to enable dynamic assessment of grasping states and real-time adjustment of control commands. For example, in a flexible grasping system based on visual detection and force-sensing feedback, the system uses a multi-channel neural network to fuse multimodal data, enabling real-time regulation of gripping force. This effectively prevents grasping failures or object damage, thereby improving grasping success rates and operational smoothness [7].

In the field of human-machine collaboration, flexible robotic arms need to have the ability to comprehensively perceive and respond to human language, movements, and environmental conditions. In this scenario, multimodal fusion control integrates speech recognition, gesture tracking, 3D vision, and pose perception to accurately identify human intent and make safe decisions. For example, in certain industrial collaborative systems, a control architecture that fuses voice and visual commands can automatically generate corresponding trajectories or actions based on the language commands and gesture guidance issued by workers, while combining IMU and visual positioning for dynamic obstacle avoidance [8]. Such systems significantly enhance human-machine collaboration efficiency and the naturalness of interaction, demonstrating the powerful potential of multimodal approaches in semantic understanding and behavior generation.

In medical and rehabilitation settings, flexible robots are often used for assisted training, surgical support, and patient interaction. Such applications typically require extremely high precision in control and human perception. Multimodal perception systems combine visual tracking, electromyographic signal recognition, and haptic feedback in such settings to comprehensively model patients' movement trends and physiological states. For example, a certain flexible rehabilitation training system identifies the user's intended movements through electromyographic signal recognition, combines visual monitoring of limb movement status with haptic sensor feedback on contact force, and implements a mechanism for autonomously adjusting training paths and force control. This ensures operational safety while enhancing training effectiveness and human-machine interaction quality [9].

Path planning and obstacle avoidance in dynamic environments are also important application areas for multimodal fusion control. In complex operational scenarios, flexible robotic arms typically need to respond in real-time to changes in external obstacles, perform multi-objective path planning, and handle dynamic human interference. By fusing inputs from multiple sensors such as lidar, RGB-D cameras, infrared sensors, and IMUs, the system can perceive the environmental state in real-time and construct a dynamic map [10]. Such systems are typically equipped with graph neural networks or attention-based fusion modules to achieve semantic understanding and path re-planning across modalities. At the execution layer, they combine model predictive control (MPC) or robust control algorithms to rapidly respond to path deviations and sudden obstacles.

In addition to the aforementioned typical applications, multimodal fusion control also demonstrates significant potential in emerging scenarios such as agriculture, disaster response, and space operations. For example, in agricultural harvesting robots, by fusing visual, force, and olfactory perception, it is possible to identify fruit ripeness and adjust grasping strategies; in disaster rescue missions, by fusing infrared, visual, and audio signals, personnel search and dynamic navigation can be achieved; in space robotic arm tasks, multimodal information fusion is used to enhance the stability and fault tolerance of control systems under extreme conditions.

In summary, multimodal fusion control has gradually transitioned from laboratory algorithms to actual system deployment, finding widespread application in multiple critical task scenarios involving flexible robotic arms. By fully leveraging the complementary nature of each modality's perception capabilities, it constructs highly robust and adaptive control systems, providing strong support for precise operations, human-robot collaboration, and intelligent interaction in uncertain environments. With ongoing advancements in hardware integration and algorithm evolution, this technology is poised to expand further into more complex intelligent systems characterized by enhanced human-robot collaboration in the future.

Multimodal fusion control enables flexible robotic arms to perform intelligent operations in complex tasks and unstructured environments. However, during its systematic deployment and large-scale application, it still faces numerous practical technical challenges. These challenges span multiple aspects, including the processing of perception data, coupling between modalities, adaptation of fusion strategies, and the overall interpretability and scalability of the system. Additionally, with the diversification of sensor types and the increasing complexity of tasks, future multimodal fusion control will continue to evolve toward higher adaptability, stronger robustness, and end-to-end intelligent decision-making.

First, the real-time and synchronization issues of data fusion are one of the key bottlenecks in current engineering implementation. The data sampling frequencies, communication protocols, and time delays of different modality sensors vary significantly. For example, cameras operate at 60 Hz, IMUs at up to 200 Hz, and force sensors may exceed 1000 Hz, leading to complexities in establishing a unified time reference and alignment processing within the control system. Additionally, in robotics tasks, control responses typically need to be completed within tens of milliseconds, but the temporal integration and feature fusion of multimodal information often result in fusion delays, which in turn affect control accuracy and system stability.

Second, calibration and spatial alignment issues for different sensors remain technical challenges in system integration. Visual, force, and tactile sensors are often distributed across different components or even platforms, each with distinct coordinate systems and response characteristics. High-precision calibration methods (such as external parameter estimation and dynamic registration algorithms) are required to achieve spatial alignment. Especially in flexible robotic arm structures where elastic deformation occurs, the relative positions of sensors dynamically change over time, making traditional static calibration methods inadequate. Additionally, redundant sensor deployment can lead to communication bandwidth occupancy and hardware coupling risks, increasing system complexity and maintenance costs.

At the fusion control strategy level, mechanisms for addressing modal conflicts and information redundancy also require urgent resolution. When information from different modalities conflicts semantically at the decision-making layer—for example, when visual sensors identify target A but tactile feedback indicates no contact—the system struggles to determine which modality to trust. Furthermore, multimodal systems often include redundant modalities, which can enhance fault tolerance but may also cause feature redundancy and information interference, thereby affecting fusion performance. Therefore, establishing mechanisms for modal confidence assessment, conflict detection, and dynamic weight allocation is a prerequisite for achieving a stable control system.

To address these challenges, future multimodal fusion control may advance in the following directions:

First, an end-to-end architecture that integrates perception and control. Current mainstream methods predominantly adopt a "perception-fusion-decision" staged strategy, which, while structurally clear, is prone to local optima and information fragmentation during model training and execution. Future research will increasingly adopt structures such as deep reinforcement learning, transformers, and multimodal graph neural networks to directly map raw multimodal inputs into control strategies, thereby achieving a highly synergistic end-to-end closed-loop learning system that integrates perception and control, reducing the need for manual feature design and rule-based intervention.

Second, build a generalized, multi-task adaptable multimodal perception module. Current multimodal control systems are often designed for a single task and have limited generalization capabilities. In the future, self-supervised learning, transfer learning, and modality contrast learning technologies can be used to develop perception fusion models with cross-scenario and task generality, achieving modality adaptation and feature sharing, thereby improving system deployment efficiency and environmental adaptability.

Third, improve the system's modality robustness and adaptive scheduling capabilities. Future multimodal systems should have modality "hot-swap" and modality degradation capabilities, automatically adjusting fusion strategies and control logic when a modality fails or signals are abnormal to ensure task continuity. Related technologies include modal health status assessment, modal weight adjustment mechanisms, and modal switching discriminators. Some studies have already achieved dynamic modal management through neural network confidence assessment or Bayesian inference.

Finally, with the deep integration of artificial intelligence into control systems, the issues of explainability and control transparency in modal fusion control systems are gradually gaining attention. Current deep fusion models are often regarded as "black boxes," making it difficult to trace how specific modes influence control decisions. This poses a serious risk for applications with high-safety requirements (such as healthcare and collaborative manufacturing). Therefore, future efforts should focus on enhancing the explainability design of control systems, such as introducing visualizable modal attention distributions, semantic path tracing, and decision visualization mechanisms to improve system transparency and human-machine collaboration trust.

In summary, although multimodal fusion control has demonstrated significant advantages in flexible robotic arm applications, its engineering implementation still faces a series of challenges, including data synchronization, sensor calibration, modal conflicts, and the generalization capabilities of fusion algorithms. With improvements in computational resources and learning mechanisms, this technology will continue to break through in areas such as end-to-end fusion, modal scheduling, and system explainability, becoming a key technological foundation for the next generation of intelligent flexible operating systems.

3.3 Summary

This section provides a systematic discussion on the key role and implementation pathways of multimodal fusion control in flexible robotic arms. It begins by highlighting the strong demand for multidimensional perception and intelligent control in unstructured environments, as outlined in the introduction, and then explores the composition of multimodal perception, the characteristics of typical modalities, and the necessity of perception collaboration. Building on this foundation, the section conducts an in-depth analysis of the technical layers involved in multimodal data fusion, including data-layer, feature-layer, and decision-layer fusion mechanisms, as well as the spatiotemporal alignment and information heterogeneity issues that must be addressed to implement these mechanisms.

In terms of control strategies, the paper provides a detailed discussion of typical methods for fusion control within a hierarchical architecture, including path planning strategies combining model predictive control (MPC), adaptive modal weighting mechanisms, end-to-end learning methods based on deep reinforcement learning, and robust control and degraded operation mechanisms under modal failure conditions. These strategies demonstrate the complete closed-loop process of fusion control from "perception input" to "action output" and exhibit excellent practical performance across multiple typical task scenarios.

At the application level, multimodal fusion control in flexible robotic arms has been widely applied in critical tasks such as object grasping and manipulation, human-robot collaboration, medical rehabilitation, and dynamic obstacle avoidance. Through an analysis of actual systems, the paper demonstrates the significant value of multimodal fusion in enhancing operational precision, environmental adaptability, and interaction flexibility. At the same time, several challenges faced in system deployment are identified, such as latency bottlenecks in real-time fusion, calibration and communication issues with heterogeneous sensors, the absence of a mechanism for identifying modal conflicts, and insufficient interpretability of fusion models.

Looking ahead, with the further development of technologies such as deep learning, self-supervised learning, and multimodal graph networks, multimodal fusion control will continue to evolve in terms of intelligence, adaptability, and explainability. Robotic arms based on this technology are expected to become core equipment for complex dynamic environments, high-safety collaborative scenarios, and precise task operations, driving robotic systems from "perception and understanding" toward true "cognitive control."

References

1. Botsali, F. M., Tinkir, M., Kalyoncu, M., et al. (2010). Fuzzy logic trajectory control of flexible robot manipulator with rotating prismatic joint [C]. In *2010 The 2nd International Conference on Computer and Automation Engineering (ICCAE)* (Vol. 3, pp. 35–39). IEEE.
2. Sun, C., He, W., & Hong, J. (2016). Neural network control of a flexible robotic manipulator using the lumped spring-mass model. *IEEE Transactions on Systems, Man, and Cybernetics: Systems, 47*(8), 1863–1874.
3. Er, M. J., & Gao, Y. (2003). Robust adaptive control of robot manipulators using generalized fuzzy neural networks. *IEEE Transactions on Industrial Electronics, 50*(3), 620–628.
4. Hu, S., Han, Y., Liu, M., et al. (2023). Research on motion control strategy of flexible manipulator based on swarm intelligence optimization. *IEEE Access, 11*, 139409–139421.
5. Zhou, K., Mao, B., Zhang, Y., et al. (2024). A cable-actuated soft manipulator for dexterous grasping based on deep reinforcement learning. *Advanced Intelligent Systems, 6*(10), 2400112.
6. Khoobjou, E., & Mazinan, A. H. (2017). On hybrid intelligence-based control approach with its application to flexible robot system. *Human-Centric Computing and Information Sciences, 7*(1), 5.
7. Liang, Q., Xiao, W., Long, J., et al. (2025). A multimodal robust recognition method for grasping objects with robot flexible grippers. *IEEE/ASME Transactions on Mechatronics, 30*(2), 1154–1165.
8. Duan, J., Zhuang, L., Zhang, Q., et al. (2024). Multimodal perception-fusion-control and human–robot collaboration in manufacturing: A review. *The International Journal of Advanced Manufacturing Technology, 132*(3), 1071–1093.
9. Zhou, S., Li, Y., Wang, Q., et al. (2024). Integrated actuation and sensing: Toward intelligent soft robots. *Cyborg and Bionic Systems, 5*, 105.
10. Donato, E., Falotico, E., & Thuruthel, T. G. (2024). Multi-modal perception for soft robotic interactions using generative models [C]. In *2024 IEEE 7th International Conference on Soft Robotics (RoboSoft)* (pp. 311–318).

Applications of Intelligent Flexible Manipulators

1 Manufacturing and Automation

Manufacturing industry refers to the sector that, during the era of mechanical industry, utilizes various resources (such as materials, energy, equipment, tools, capital, technology, information, and labor) to transform them—through manufacturing processes and in accordance with market demands—into large-scale tools, industrial products, and consumer goods for people's use and daily life. Manufacturing industry is one of the important industries in China, which is the main body of the national economy and an important part of the modernized economic system, and at the same time, China is also the world's first manufacturing country. In order to promote the continuous development of the manufacturing industry, the automation system of the manufacturing industry is gradually being established and optimized. The core goal of automation is to simplify the process, improve efficiency, and reduce human error. The application of robots in the manufacturing industry can bring about multiple advantages such as automated production, reduction of production costs, and improvement of product quality, while also having broad application prospects [1].

With the continuous development of robotics, more and more robots are being used in different work scenarios. Due to the use of materials with large deformation, high flexibility, and high load capacity, flexible robots are highly pliable and environmentally adaptable. Compared to traditional rigid robots, flexible robots show more potential in complex and unstructured environments. In addition, combined with intelligent and advanced control algorithms, the manipulator can achieve more precise and intelligent operation and be adapted to more demanding application scenarios. Therefore, intelligent flexible manipulators have a wide range of applications in manufacturing and automation, including but not limited to precision parts assembly, parts quality and accuracy testing, food processing, complex space intelligent operation, industrial assembly line sorting, and intelligent warehousing.

J. Qu, *Robotic Flexible Manipulation and Intelligent Sensing Technologies*,
https://doi.org/10.1007/978-981-95-6232-9_5

1.1 Automation in Agriculture

Agricultural automation is an important branch in the field of manufacturing and automation. The use of intelligent flexible manipulators to complete fruit picking, intelligent seeding, variable fertilizer application, pest management, and other tasks is an inevitable trend in the development of agricultural automation. Through the use of intelligent flexible manipulator, it can reduce the dependence of agriculture on labor costs, improve the efficiency and quality of work, and thus improve the economic benefits of agricultural production.

Among them, the use of intelligent flexible manipulator to complete crop picking is an important and representative application in the field of agricultural automation. For some fragile and vulnerable crops, if ordinary manipulators are used to pick, it will lead to part of the crop being damaged, resulting in economic losses. In contrast, firstly, the flexible material gripper used in the intelligent flexible manipulator itself can play a certain protective role for the crops, and secondly, it can control the manipulator according to intelligent control algorithms and visual detection to accurately and gently grasp the crops [2]. For example, the intelligent *Agaricus bisporus* picking device based on FES-YOLOv5s designed in Fig. 1 is mainly composed of a frame, a camera, a truss-type robotic arm, a flexible gripper, and a control system. It is based on machine vision and uses the FES-YOLOv5s deep learning target detection model to accurately identify and localize mushrooms. At the same time, the Jetson Orin Nano is adopted as the main controller and the S-curve acceleration and deceleration motor control algorithm is used to control the robotic arm and flexible gripper to complete the harvesting of *Agaricus bisporus* mushrooms. The experimental results show that the harvesting success rate of this harvesting device is 94.95%, the average damage rate is 2.67%, and the average harvesting yield rate is 87.38%. From this, we can clearly see that the intelligent flexible manipulator has two major features of intelligence and flexibility [3].

Figure 2 shows an ellipsoidal type intelligent flexible manipulator for fruit and vegetable picking. Its overall structure is mainly composed of bracket, airbag clamping claw, compression pump, solenoid valve, coding motor, electrical components,

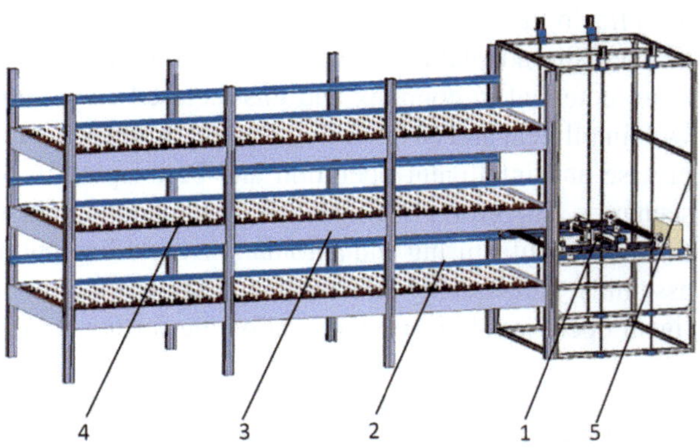

Fig. 1 Structural diagram of the *Agaricus bisporus*-harvesting platform: 1. harvesting robot; 2. U-shaped guide rail; 3. mushroom rack; 4. mushroom bed; 5. climbing device [3]

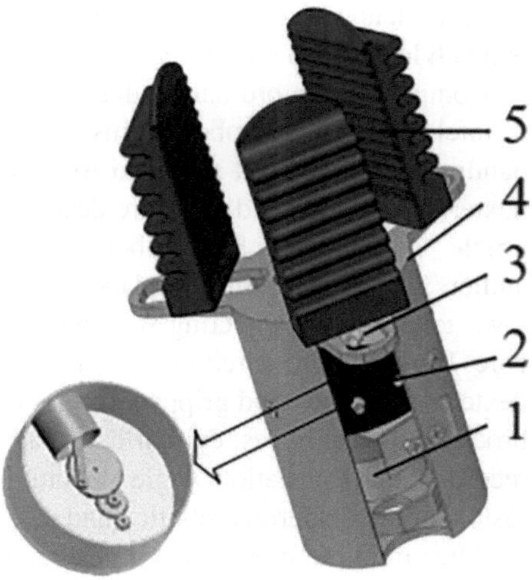

Fig. 2 The whole structure of the picker: 1. Coded motor, 2. compression pump, 3. bracket, 4. solenoid valve, and 5. airbag clip [4]

and so on. Its working principle is to first judge the type of the picking object through the visual recognition module; then to call up the preset optimal clamping force of the type in the database; then to flexibly inflate the airbag clamps to reach the appropriate degree of pressure; and finally to control the picking hand to pick the crop. On the basis of intelligence, the airbag clip is made of silicone, a flexible material that can produce large deformation. After testing, it can be seen that the manipulator can be adapted to a variety of different quality, growth state of fruits and vegetables, picking success rate of 92.5% or more, clip damage rate is less than 7.5% [4].

However, intelligent flexible manipulators are still facing some challenges in crop picking, for example, facing the complex and changing natural environment, factors such as light, branch, and leaf shading will affect the recognition and positioning accuracy of the manipulator; secondly, the picking efficiency of the manipulator is not high enough, and further optimization of the algorithms and mechanical structure is needed; in addition, the cost of manipulator research and development and manufacturing needs to be lowered but at the same time, the performance should also be maintained.

1.2 Automation of Plant Applications

The automation of factory production lines has also received widespread attention and emphasis from people. An increasing number of enterprises have put forward the concepts of unmanned factories and smart factories, and intelligent flexible manipulators such as robotic arms play an important role in this context. To reduce labor costs and improve work efficiency, many factories choose to use industrial robotic arms for assembly line work. The main reason is actually that robotic arms can complete a large number of repetitive tasks for a long time, and their error rate is much lower than that of manual operations.

Compared with ordinary industrial robots, first of all, the flexible end-effectors of intelligent flexible robotic arms can adapt to a variety of different assembly and handling tasks without the need to switch effectors repeatedly. For example, the dexterous robotic hand structure designed for delicate in-hand manipulation in the article "A Dexterous Soft Robotic Hand for Delicate In-Hand Manipulation" by authors such as Sylvain Abondance is mainly made by using two side-by-side bellows actuators and injecting silicone elastomer into a 3D-printed mold with a soft core. During the research process, the authors conducted experiments in which the dexterous robotic hand gripped different objects, including cakes of different sizes, broccoli, square boxes, cylindrical small bowls, and so on. At the same time, they recorded the translation angle deviation, rotation distance, and angle deviation results of the dexterous robotic hand during the test.

After multiple experimental verifications, Sylvain Abondance and others have proven that this intelligent flexible robotic hand can perform stable in-hand manipulation of small objects without knowing their exact position, shape, or size [5].

Secondly, the operation accuracy of the manipulator can be improved and the error rate can be reduced by some algorithms such as path optimization as well as image analysis. For example, some researchers have proposed a fuzzy neural network-based intelligent control method for manipulators with the aim of solving the precise trajectory tracking of the robotic arm when dealing with dynamic uncertainty problems. As shown in Table 1, the method is compared with PD control and also neural network control technology, and it can be seen that the fuzzy neural network-based control method trajectory tracking has a lower error rate and shorter vibration instability time, which reflects the advantages of the control method [6].

In addition, the factory regularly monitors the working status of the intelligent flexible manipulator and can fault localize it. For example, Fig. 3 presents a multi-condition fault localization algorithm for manipulators, which shows the entire troubleshooting process. It uses reinforcement learning techniques to train the debugging experience of the model in detecting parameters under different working conditions and combines the robot dynamics to realize fault detection and fault source localization in a flexible environment. After experimental simulation and testing in real industrial applications, the results show that the algorithm's fault

Table 1 Comparison of PD, neural network, and fuzzy neural network control

Methods	Link	Max. tacking error	Steady-state error	Oscillation time
PD control	Joint 1	0.0212 rad	0.0099 rad	6.42 s
	Joint 2	0.0436 rad	0.0028 rad	6.36 s
NN control	Joint 1	0.0045 rad	0.0045 rad	0.18 s
	Joint 2	0.0015 rad	0.0008 rad	0.12 s
FNN control	**Joint 1**	**0.0011** rad	**0.0011** rad	**0.12** s
	Joint 2	**0.0008** rad	**0.0002** rad	**0.06** s

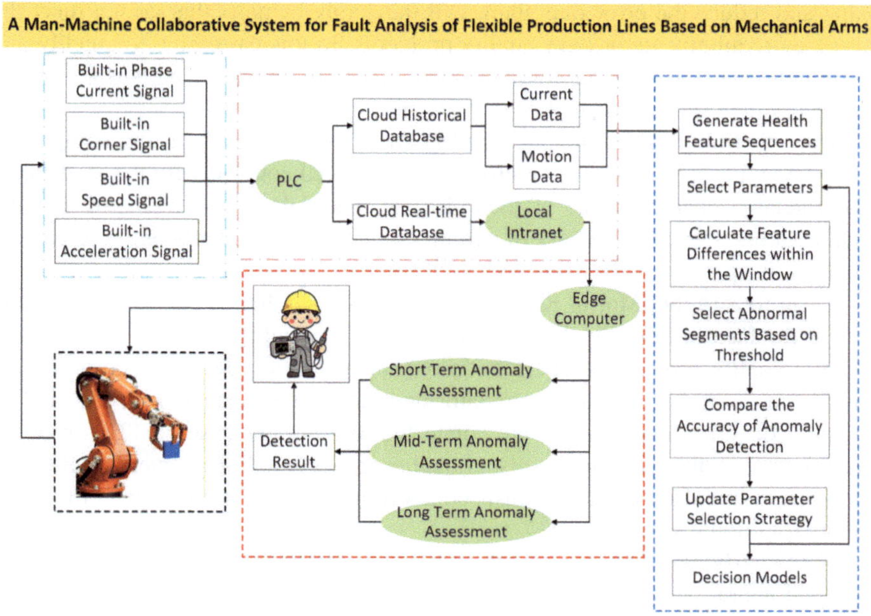

Fig. 3 The process of production line fault detection [7]

detection accuracy reaches 86% and the localization accuracy reaches 81.35%, which has a good application prospect [7].

At present, the intelligent flexible manipulator for factory applications also has to deal with some challenges, such as its sensors and control algorithms should be accurate enough to face the complex dynamic environment of the factory, to ensure that the manipulator can quickly and accurately adapt to the operation of the factory assembly line; Secondly, the flexible manipulator should be able to adjust its stiffness to match the different tasks and environmental conditions of a variety of work; In addition, the factory environment is subject to higher noise levels and numerous sources of interference. In addition, the factory environment is noisy with many disturbing factors, so the noise anti-interference ability of the intelligent flexible manipulator is required to be higher; finally, the work in the human-machine collaborative environment needs to ensure the personal safety of the operators, so we should pay extra attention to the safety of the manipulator.

1.3 Automation of Parts Quality Inspection

In the field of intelligent manufacturing and automated production, large batches of different workpieces need to be produced in an assembly line, and the parts produced in factories generally have corresponding accuracy requirements. Although it is said that industrial robots instead of manual labor can improve the accuracy of parts and reduce the error rate, there will inevitably be parts that do not meet the accuracy requirements or have quality problems. In this case, how to quickly and accurately detect and screen out poor-quality parts among many parts is a problem that every factory is trying to solve.

Thus, flexible robots integrated with a variety of advanced testing equipment, furnished with high-precision control algorithms, and coupled with the stretchable and rotatable characteristics of their flexible materials allow intelligent flexible manipulators to excel at the task of part quality inspection and efficiently screen out all defective parts.

As an illustration, Martin Kefer and Jihuan Tian, in their article "An Intelligent Robot for Flexible Quality Inspection," present an intelligent robot designed for flexible quality inspection. The robot's overall inspection device comprises three on-site units—a mechanical manipulator unit, an image enhancement unit, and an image acquisition unit—which are connected to a computer server. This robot is capable of not only conducting visual quality inspections for geometric defects in workpieces but also performing inspections from angles that are beyond the reach of any traditional sensors.

The underlying principle involves utilizing a vision system that incorporates an artificial neural network approach to carry out quality inspections on workpieces. This system can intelligently analyze the appearance quality of parts and achieve a consistent inspection time of less than 1 s for workpieces with varying levels of detail. Experimental validation has demonstrated that this method delivers superior performance when traditional template-based solutions cease to be effective [8].

Another research team designed a 3D vision system based on a motion recovery structure SFM. By having an intelligent industrial robot carrying an industrial machine vision camera scan, an object as shown in Fig. 4, an accurate 3D model of

Fig. 4 Industrial machine vision camera mounted on a KUKA AGILUS end-effector [9]

the object is generated in a computer. The method was realized by the team by integrating an automated vision system and a specially developed graphical user interface with the robotic work cell. The researchers also compared four different tool paths, namely polar coordinates, hemispherical helix, parabolic helix, and parabolic helix for some areas and finally selected the parabolic helix path for optimal image acquisition. It is verified that the parabolic helix path can be used to inspect samples with 80% success rate of capturing ultrasonic pulse-echo data, which meets the conditions of most applications [9].

In addition, a team has developed an automated optical inspection system AOI for reflective parts, which specializes in the inspection of parts made of high-pressure die-cast aluminum used in the automotive industry. The overall structure of the inspection system is shown in Fig. 5, which consists of a camera, an overhead light, a universal robot, and an intelligent manipulator. The system achieves uniform illumination by means of an overhead light so that the camera is able to capture minute defects on the reflective surface. Meanwhile, the application of deep learning algorithms allows the intelligent flexible manipulator to accurately locate and repetitively inspect parts during the inspection process, which improves the efficiency and effectiveness of the inspection process. After testing, the detection time of the system was within an acceptable operating range, proving the feasibility of

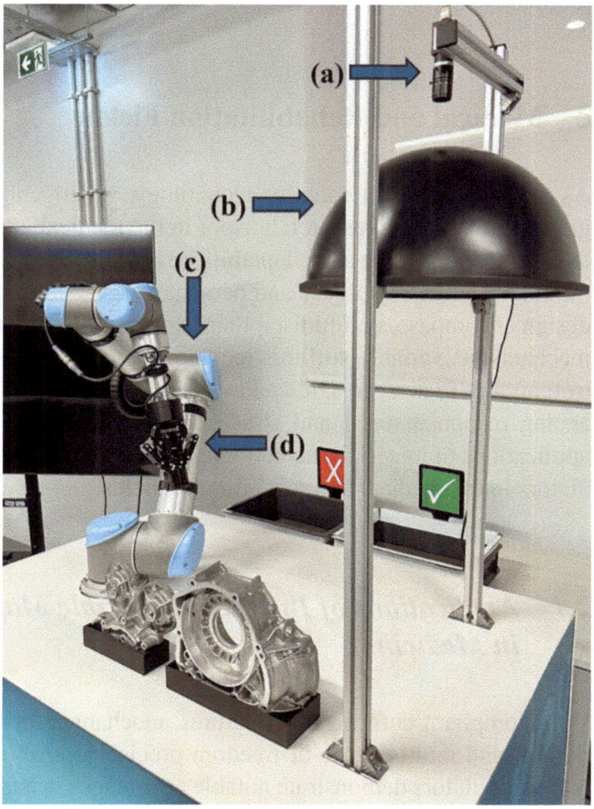

Fig. 5 AOI system setup: a. MAKO G-503B camera, b. VL-D430W dome light, c. Universal Robot UR5, d. Robotiq 2F-85 manipulator [10]

the system for real-time defect detection. In addition, they used the sliding window method (YOLOv8-Mosaic) to optimize the system in certain ways, which improved the recall of the inspection system (84.71%) while also reducing the inference time [10].

Of course, there are also areas that need to be improved in terms of parts quality detection for intelligent flexible manipulators. First, we must continuously enhance the detection equipment and visual analysis algorithms to ensure they possess sufficiently high discriminatory precision, thereby minimizing the occurrence of missed detections for any substandard workpieces. Secondly, its quality detection speed to meet the real-time operation of the assembly line requirements cannot affect the normal production of the assembly line; In addition, we can design better control algorithms to achieve fast and accurate screening of poor-quality parts to be separated out.

The above three types of examples are only a small part of the many applications of intelligent flexible manipulator in the field of manufacturing and automation, I believe that the intelligent flexible manipulator in the future in the field will have more participation in the future industrial development to play an increasingly important role. I also believe that through continuous technological innovation and application expansion, intelligent flexible manipulators will bring higher efficiency, better quality, and stronger competitiveness to the manufacturing and automation fields.

2 Medical and Rehabilitation Fields

As an emerging soft robotics technology, intelligent flexible manipulators have gradually become a research hotspot in the medical and rehabilitation fields due to their high compliance and adaptability, showing significant advantages especially in minimally invasive surgery and personalized rehabilitation assistive devices. Their design encompasses multidisciplinary integration from material selection, actuation mechanisms, variable-stiffness techniques, to flexible sensing and intelligent control, aiming to achieve safe and efficient interaction with human tissues while balancing response speed and structural reliability. This section will introduce the applications of materials, actuation, and system design of intelligent flexible manipulators in the medical and rehabilitation fields.

2.1 Applications of Intelligent Flexible Manipulators in Medicine

With compliant deformable structures, mechanical properties matching biological tissues, and multi-degree-of-freedom precise control capabilities, intelligent flexible manipulators demonstrate notable advantages in endoscopy and minimally invasive surgery. They can navigate flexibly and perform delicate operations in narrow, curved body cavities, reducing interference with healthy tissue. Combined with real-time sensing and force feedback control, the technology has been applied in clinical scenarios such as early gastrointestinal cancer resection, respiratory lesion removal, and cardiovascular intervention, improving surgical precision and safety while shortening patient recovery time. Regarding flexible actuators, Wang [11] and colleagues systematically reviewed the latest advances of soft robotic devices in the medical field. The authors described applications of flexible sensors and implantable devices in electrocardiography, neural and chemical monitoring, and focused on soft surgical instruments represented by variable-stiffness soft endoscopes and catheters, which are capable of navigating and stably positioning within narrow channels to perform biopsies, ablations, and other procedures; in rehabilitation and assistance, soft gloves, exoskeletons, and prostheses improve motor function in stroke patients with advantages of light weight and low noise.

Kong [12] et al. reviewed the research progress of magnetically controlled medical microrobots in minimally invasive therapy. The article focuses on magnetic guidewires/catheters and low-melting-point alloy (LMPA) variable-stiffness techniques, achieving second-scale soft-to-hard switching via liquid–solid phase change, allowing instruments to compliantly traverse vessels while providing rigid support at lesions. Finally, the authors list typical applications such as targeted drug delivery, PCI vascular intervention, atrial fibrillation ablation, neural navigation, and in vivo bioprinting, demonstrating the great potential of magnetically controlled microrobots to reduce radiation and improve surgical precision. Fu [13] et al. developed a magnetic guidewire robotic system MCGRS for vascular interventional surgery, established MG steering kinematics using constant curvature theory, predicted magnetic tip deformation with a dipole–Cosserat-rod coupled model, and designed a trajectory planning algorithm to achieve automatic navigation. Zhang [14] et al. from Zhejiang University designed an AI-assisted bronchoscopy robot; the system hardware uses a magnetically attached quick-change catheter capable of reaching the ninth-generation bronchi and a four-line-driven flexible distal end, while on the software side it implements an AI–human shared control algorithm: the policy network takes real-time bronchoscope images and discrete doctor commands as input and outputs continuous steering actions to keep the scope centered in the lumen. In [15], Dreyfus et al. reported a highly dexterous hinged continuum robot controlled by magnetic forces; it overcomes pushability limits through rotation to improve catheter navigation in tortuous vessels. The device successfully navigated intravascularly in vivo from the aorta to millimeter-scale cerebral arteries, demonstrating its future potential for nontraumatic access to occluded vessels in the brain.

In [16], the authors proposed an origami–pneumatic-structure-based soft neurosurgical retractor to replace traditional rigid metal instruments. After insertion through a cranial burr hole under 15 mm in diameter, the device can expand to a 25 mm diameter by ±40 kPa pressure control, creating a minimally traumatic surgical channel to deep lesions. Compared with traditional stainless-steel retractors, the retractor designed in the paper applies 4.8 times less force to a brain model, significantly reducing the risk of tissue damage. In [17], the authors combined the advantages of concentric tube continuum robots and notched continuum robots to design a novel bio-inspired composite continuum robot suitable for minimally invasive surgery. In [18], the authors proposed integrating a fiber-blocking variable-stiffness mechanism into the STIFF-FLOP soft surgical arm to balance flexibility and stiffness. By comparing two new modules: Type A places fibers in the central channel, preserving tri-directional bending but sacrificing a tool channel; Type B replaces two pairs of drive chambers with fiber-blocking chambers, retaining a 4.5 mm channel but allowing only unidirectional bending. The study shows that fiber blocking can effectively provide reversible stiffening in slender soft structures, offering a new route for minimally invasive surgical instruments to progress from flexible exploration to stable operation.

Mitros [19] et al. designed a novel continuum robot bronchoscope for distal lung sampling of mechanically ventilated patients in the intensive care unit. This robot has 7 degrees of freedom, capable of dexterous manipulation and visualization of the distal airway. The authors also proposed a new numerical solver to improve the model's computational efficiency, which is able to predict the robots' shape within less than 0.011 s, lying the foundation for future deployment of robotic bronchoscopes in mechanically ventilated patients. Duan [20] et al. proposed an ultrathin robotic bronchoscope system that integrates electromagnetic navigation with variable-stiffness and variable-length bending segments, which is for minimally invasive biopsy of pulmonary nodules via natural cavities. The system integrates EM tracking with centerline-based automatic path planning, enabling preoperative 3D reconstruction, intraoperative real-time registration, and master-slave remote control. In vitro airway model testing demonstrated its high flexibility, environmental adaptability, and navigation accuracy (Fig. 6).

2.2 *Application of Intelligent Soft Manipulators in Rehabilitation*

Up to 75% of stroke survivors are left with motor impairments, making everyday actions such as holding a cup, eating, or buttoning impossible and severely limiting the ability to perform daily activities. Traditional rigid exoskeletons are difficult to wear long-term due to weight, alignment errors, and safety concerns. Intelligent soft manipulators, typically built with fabric-silicone composite structures and driven pneumatically or by SMA, are lightweight, low noise, and skin-friendly and breathable, allowing patients to perform imperceptible continuous training at home, in the office, or outdoors, significantly enhancing neural plasticity and functional transfer. Soft manipulators have thus become a key bridge linking hospital, community, and home rehabilitation chains, with profound implications for reducing caregiver burden and lowering societal healthcare costs.

Panagiotis [21] et al. at Harvard University proposed a hydraulically driven soft rehabilitation glove specifically designed for hand dysfunction in stroke and spinal cord injury patients. The fiber-reinforced elastomer actuators with rectangular

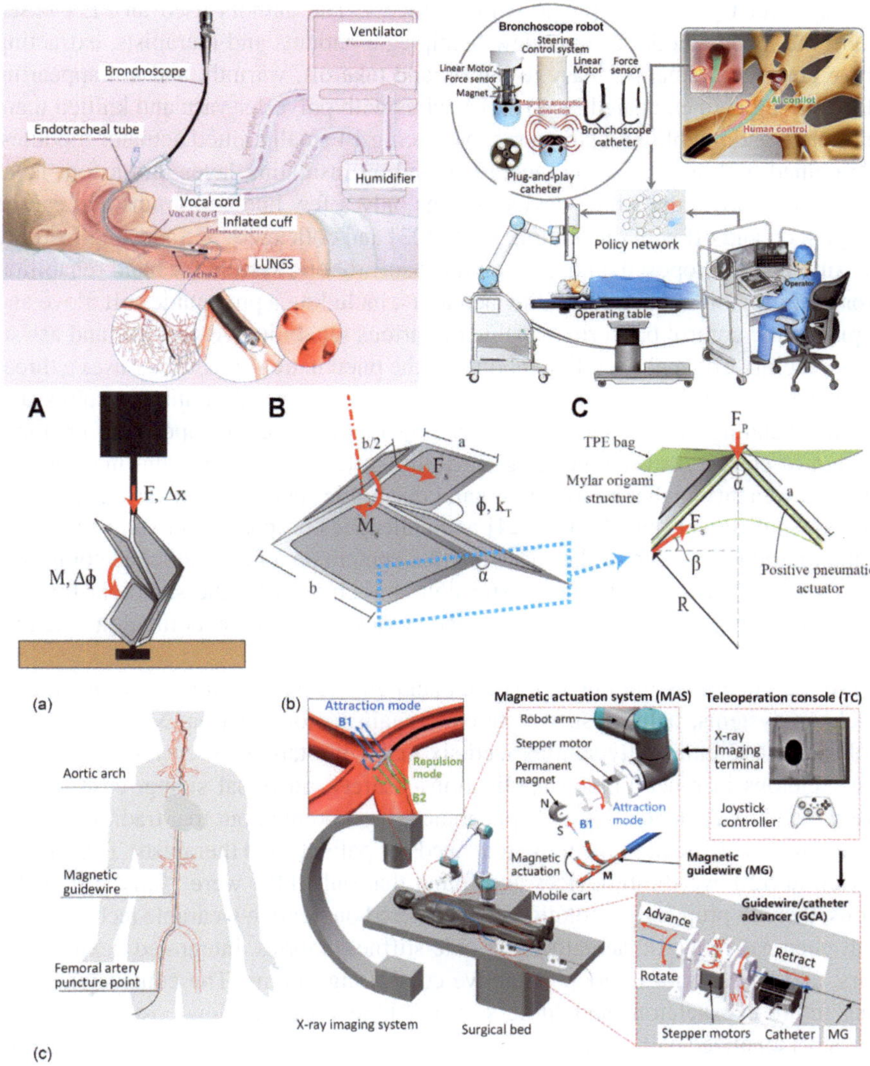

Fig. 6 Application of intelligent flexible actuators in the medical field

cross-sections are only 7 mm thick, yet when pressurized can perform combined bending-extension-twisting motions, and with just two segmented configurations can cover the full grasping trajectories of all five fingers. The glove retains only a textile fixation layer and soft actuators at the extremity, with a total weight of <0.5 kg, enabling task-oriented training in home or clinic environments. Compared with traditional rigid exoskeletons, this solution balances safety, portability, and low cost, offering a novel rehabilitation pathway for hand dysfunction after stroke or spinal cord injury. Lee [22] and others made shape memory alloy (SMA) the first soft actuator embedded into a rehabilitation glove, providing a truly daily-wearable soft exoskeleton for hemiplegic stroke patients. The authors used an FEA-based consumer needs model to interview hemiplegic patients and therapists, extracting three core requirements: "easy to put on and take off, warmth, and not appearing disabled"; they then coated Ni-Ti SMA wires with polyester yarn and knitted them into plain and double-knit structures, producing 14 small knitted actuator modules. Combined with elastic fabric, neoprene, and an insulating layer into a four-layer open-palm glove, the whole device weighs only a few hundred grams and can be toggled on and off with one button. Wang [23] and colleagues at Shenyang Institute of Automation proposed a novel pneumatic soft robotic system for hand rehabilitation featuring bidirectional bending actuators, including a pneumatic soft glove and a pneumatic control platform, capable of various rehabilitative gestures and assisting grasping. The main bending module of the pneumatic soft actuator uses a three-chamber structure, allowing a wider range of finger rehabilitation gestures and greater bending angles; the reverse-bending module uses a trapezoidal chamber design to enhance reverse-bending capability, effectively promoting finger extension movements so that people with hand disabilities can perform various gestures and grasp different objects. Yap [24] and colleagues proposed and validated a soft rehabilitation glove for stroke survivors that centers on a fabric-silicone composite pneumatic actuator; flexion and extension are achieved by the same elastic base fabric, enabling daily grasp assistance and possessing home rehabilitation potential (Fig. 7).

At the same time, intelligent flexible manipulators play an important role in exoskeleton systems, helping stroke patients regain motor function. Morris [25] and colleagues systematically summarized six major limitations of current lower-limb exoskeletons in rehabilitation applications—safety, universal sizing, ease of use, weight and placement, cost, and appearance—and pointed out that traditional rigid structures are difficult to meet the real needs of patients and therapists. The authors reviewed 52 lower-limb devices and found that only 11% were "fully soft" solutions, thereby proposing a solution path centered on flexible wearable technologies: soft pneumatic/electric actuators, variable-stiffness fabrics, integrated flexible sensors, and "human-in-the-loop" adaptive control algorithms. The conclusion states that future exoskeletons should be soft, machine-washable, low-cost, and able to learn personalized gait in real time so as to truly integrate into home and community settings. In [26], the authors focused on the ReStore soft exosuit and conducted 5 weeks of high-intensity robot-assisted gait training (ROB)—15 sessions—with 30 subacute stroke patients, compared with standard therapy (STT); the ROB group was significantly better on the primary outcome mRS, demonstrating that soft exosuits can significantly accelerate recovery of gait, balance, and functional independence in early stroke rehabilitation, with benefits that continued to accumulate after the training ended and surpassed traditional interventions. Kim [27] and colleagues developed a soft, fully portable, lightweight exosuit that can assist hip extension for each step to reduce the metabolic cost of running and walking. A portable exosuit that assists hip extension can reduce metabolic cost by 9.3% for walking on a treadmill at 1.5 m/s and by 4.0% for running at 2.5 m/s. These reductions are comparable to carrying 7.4 kg and 5.7 kg less weight during walking and running, respectively.

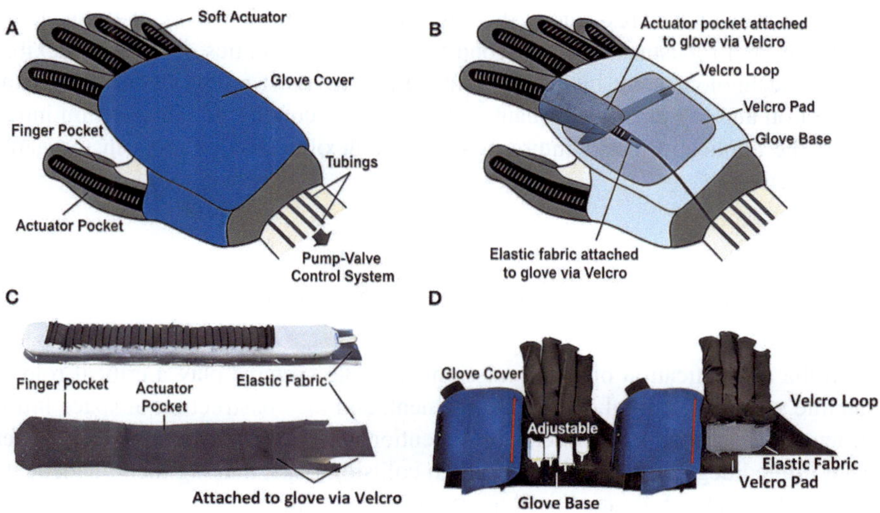

Fig. 7 Soft manipulator based on intelligent flexible actuator

In [28], the authors reviewed "pneumatic exoskeleton" technologies used for upper-limb orthopedic rehabilitation, and around the three core demands of "lightweight, low cost, easy to wear," they analyzed the evolution over more than a decade of various prototypes from bulky rigid designs to lightweight soft systems. Pneumatic actuation performs best in the trade-off between force-to-weight ratio and safety; through comparisons of joint alignment, actuation, and customization strategies across more than ten prototypes, the authors pointed out that a "minimalist pure-pneumatic" approach remains the optimal route to reduce weight, cost, and failure. Lin [29] and colleagues at Harbin Institute of Technology systematically reviewed 45 studies since 2020 on walking-assist exoskeletons, categorizing them into rigid, flexible, and tethered platforms; they noted that the hip and ankle are key targets for assistance, that tethered platforms significantly improve metabolic benefits via weight reduction and personalized control strategies, and that future work must overcome challenges in joint alignment, lightweight actuation, and human–machine coordination to expand rehabilitation and industrial applications.

3 Aerospace Field

In the aerospace domain, space robots have been deployed for extravehicular operations and deep-space exploration, significantly reducing mission risks and costs. Conventional space robots are primarily designed with rigid structures, offering limited degrees of freedom that restrict their operational capability; moreover, their weak compliance makes them unsuitable for tasks requiring soft interaction. In contrast, soft robotic systems, by virtue of their intrinsic compliance, exhibit unique actuation and impact resistance, thereby mitigating mission risks and enhancing flexibility and robustness [30]. Beyond flexible space robotics, intelligent flexible sensing technologies are increasingly integrated to assist robots in environmental perception and to monitor astronauts' physiological conditions. This section introduces the applications and future potential of flexible manipulators in the aerospace sector.

3.1 Flexible Assembly and On-Orbit Servicing

With the diversification of space missions, flexible systems play a critical role in satellite servicing, orbital debris management, and the construction of space infrastructure. Regarding end-effectors, conventional rigid robotic arms often suffer from limited degrees of freedom and high collision risks, making them inadequate for the demands of future missions requiring high flexibility and reliability. To address these challenges, researchers have proposed manipulators based on soft materials. For example, Hammond et al. designed a soft end-effector composed of polyurethane flexible structures and layered adhesive materials, where heat-activated bonding layers enable precise adhesion and separation of components. This device can operate in confined spaces with minimal impact, thereby reducing the risk of damage to spacecraft platforms [31]. In on-orbit servicing (OOS) missions, unmanned spacecraft equipped with robotic arms is required to capture and manipulate non-cooperative targets to ensure mission success. Space debris, in particular, poses a critical threat, as the reliable capture and removal of non-cooperative objects are essential to avoid generating additional debris. Ruiz et al. [32] proposed a thermally engineered multilayer flexible gripper that integrates TPU, silicone rubber, PTFE, and aerogel to withstand extreme temperature fluctuations; experiments demonstrated reliable adaptability from −196 °C to high-temperature environments. Besides thermal challenges, the vacuum conditions in space render conventional pneumatic grippers ineffective, while opposable palm-type grippers are unsuitable for large, smooth debris. To overcome this, Jiang et al. [33] proposed a dry-adhesion-based robotic gripper capable of delicately grasping, manipulating, and releasing flat or curved non-cooperative objects up to one meter in diameter under microgravity conditions. Inspired by the bistable mechanism and energy release strategy of the Venus flytrap, Zhang et al. [34] designed a bistable origami-inspired gripper actuated by shape memory alloys. This system can achieve effective capture within approximately 300 ms. In space infrastructure, solar panels must be sufficiently large to provide adequate power for spacecraft. Due to the limited internal space of CubeSats, solar panels must be foldable for compact storage and deployment. As early as the 1990s, Guest and co-workers presented a complete prototype for solar panel design based on the "flasher" concept and the zero-thickness assumption [35].

3.2 Flexible Robots for Aircraft and Planetary Exploration

Soft robotics has also demonstrated significant potential in planetary exploration and aerospace vehicle applications. IEEE Spectrum reported on NASA's "PUFFER" robot, a planar, foldable origami-inspired explorer capable of being stowed on Mars rovers and later deployed to explore hazardous terrains such as deep sand dunes, crater walls, and confined crevices. To address the challenges of uneven and unpredictable environments, tensegrity structures—composed of compression rods and tensile cables—have attracted attention in soft robotics. Under equilibrium conditions, the rods and cables experience pure compression and tension, respectively, enabling passive force distribution throughout the structure. This property provides enhanced resilience to impact, making tensegrity systems robust mobile platforms for next-generation landers and rovers. An example is NASA Ames Research Center's "SUPERball," a spherical underactuated tensegrity planetary exploration robot. Its locomotion is driven by actuating the structural cables, enabling rolling motion as its geometry deforms [36]. Similarly, the TT-3 tensegrity robot, developed jointly by the University of California, Berkeley and NASA Ames Research Center, is based on a spherical six-rod tensegrity structure. It features a unique rod-centered modular distributed actuation and control architecture, resulting in a lightweight, low-cost, and modular robotic system capable of rapid prototyping [37].

More recently, a new aerial platform—the aerial continuum robot—has been introduced by combining multirotor UAVs with flexible, trunk-like arms. Peng et al. [38], as reported in Nature Communications, developed an aerial continuum robotic system that integrates a tendon-driven flexible arm onto a small quadrotor. This configuration allows navigation through narrow slits and pipelines, as well as the manipulation of slender, deformable, or heavy objects. Experiments demonstrated successful operations such as pipe traversal and grasping of targets with varying geometries, showing strong potential for tasks like high-altitude power line inspection and pipeline maintenance.

In summary, flexible structures and soft actuation technologies have provided aerospace robotics with greater adaptability and autonomy, enabling exploration in constrained and unknown environments. The integration of flexible manipulators with aerial platforms broadens the operational range beyond that achievable by traditional rigid robotic arms.

3.3 Flexible Materials and Intelligent Sensing

Aerospace missions impose extreme demands on materials and sensors: fluid behavior under microgravity deviates from terrestrial norms, the vacuum of space induces thermal shrinkage and deformation, and solar irradiation and magnetic radiation challenge material durability. Consequently, research on novel flexible materials—such as gallium-based liquid metals, polymer gels, and electroactive polymers

(EAPs)—and integrated sensing systems has gained momentum in aerospace contexts. These materials exhibit desirable stretchability, conductivity, and adaptive deformation characteristics but must also demonstrate low-temperature resilience and minimal outgassing. For instance, Ras Labs' synthetic muscle EAPs underwent a year-long radiation exposure experiment aboard the International Space Station, successfully withstanding temperatures from −271 to 135 °C and high radiation levels. Robotics researchers are exploring their potential to replace belts, pulleys, gears, or motors for actuation and control.

With the widespread application of soft robotics and spaceborne devices, the demand for tactile and force sensing continues to increase. Jahanshahi et al. [39], in a 2025 review published in *Chinese Journal of Aeronautics*, discussed tactile sensing technologies designed for harsh space environments, evaluating resistive, capacitive, piezoelectric, and optical sensors under microgravity, vacuum, and extreme thermal conditions. For example, resistive pressure sensors are simple and robust, functioning across temperature extremes, albeit with limited sensitivity; optical sensors, by contrast, offer strong electromagnetic interference resistance and high precision in distance measurement.

For the long-term sustainability of crewed space missions, real-time monitoring of astronauts' physiological and psychological health is essential. Unique space conditions—including microgravity, cumulative radiation exposure, and extreme thermal cycling—pose risks such as bone density loss and muscle atrophy. Researchers have proposed integrating flexible electronics into astronaut health monitoring and environmental sensing. Flexible wearable patches, for example, can conform to the skin and track vital signs such as heart rate and oxygen saturation in real time, provided they are radiation- and low-temperature-resistant to ensure signal stability. Wang et al. [40], in Wearable Electronics, reviewed flexible wearable technologies in aerospace medicine, emphasizing multimodal arrays that combine inertial measurement units, bioelectrodes, and environmental sensors to monitor cardiopulmonary, neuromuscular, and circadian signals for early health anomaly detection.

In aeronautics, though operating conditions are less harsh than those in space, flexible sensing systems find broader applications. For instance, Soriano-Amat et al. [41], in Sensors, reported the monitoring of highly flexible aircraft model wings using distributed fiber-optic sensors based on time-domain extended Φ-OTDR. This system detected wing bending, torsion, and vibration with centimeter-level spatial resolution and megahertz-level bandwidth in real time. By embedding standard optical fibers within wing samples, the researchers reconstructed deformation profiles under various load conditions, demonstrating the feasibility and efficiency of fiber-optic sensing in monitoring flexible aerospace structures.

4 Deep-Sea Exploration Field

Intelligent flexible manipulators demonstrate significant application potential in deep-sea exploration, primarily in biological sampling, minerals resource collection, and underwater equipment inspection and maintenance. Leveraging their material properties and structural design, flexible manipulators can perform precise operations in complex and variable deep-sea environments, enhancing operational efficiency and accuracy. [42] The following sections detail their applications in these three domains.

4.1 Deep-Sea Biological Sampling

Deep-sea biological sampling constitutes a fundamental task in deep-sea exploration. Given the fragility of organisms in these environments, traditional rigid sampling equipment may damage biological specimens, compromising research accuracy. Intelligent flexible manipulators, through flexible materials and biomimetic design, effectively enhance sampling success rates and specimen integrity.

1. **Flexible Materials and Bionic Structural Design**
   Smart flexible manipulators utilize flexible materials such as silicone and elastomers, which are deformable and impact-resistant. This allows them to adapt to biological specimens of varying shapes, preventing damage from excessive compression during capture. Bionic design integrates natural biological movement characteristics—such as octopus tentacles or starfish arms—into the manipulator's structure, enabling multi-degree-of-freedom operation. [43] This biomimetic architecture not only enhances grasping flexibility but also improves adaptability to irregular organisms like mollusks or branching corals (Fig. 8).

2. **Non-destructive Collection of Biological Samples**
   A key advantage of flexible manipulators lies in their ability to achieve non-destructive sample collection through precise force control. For instance, a flexible robotic arm developed by the Woods Hole Oceanographic Institution can collect various mollusks and corals at depths up to 4000 meters while preserving their biological viability. This device utilizes force feedback control technology to precisely regulate applied forces during grasping, preventing physical damage to samples under high-pressure conditions. [44] Furthermore, flexible manipulators can integrate sensors to enable real-time perception of target organisms' mechanical properties (such as hardness and flexibility), further optimizing grasping strategies (Fig. 9).

3. **Enhanced Sampling Efficiency and Sample Integrity**
   Compared to rigid manipulators, the adaptive capabilities of intelligent flexible manipulators significantly improve deep-sea biological sampling efficiency and sample integrity. The multi-degree-of-freedom adjustment of flexible structures

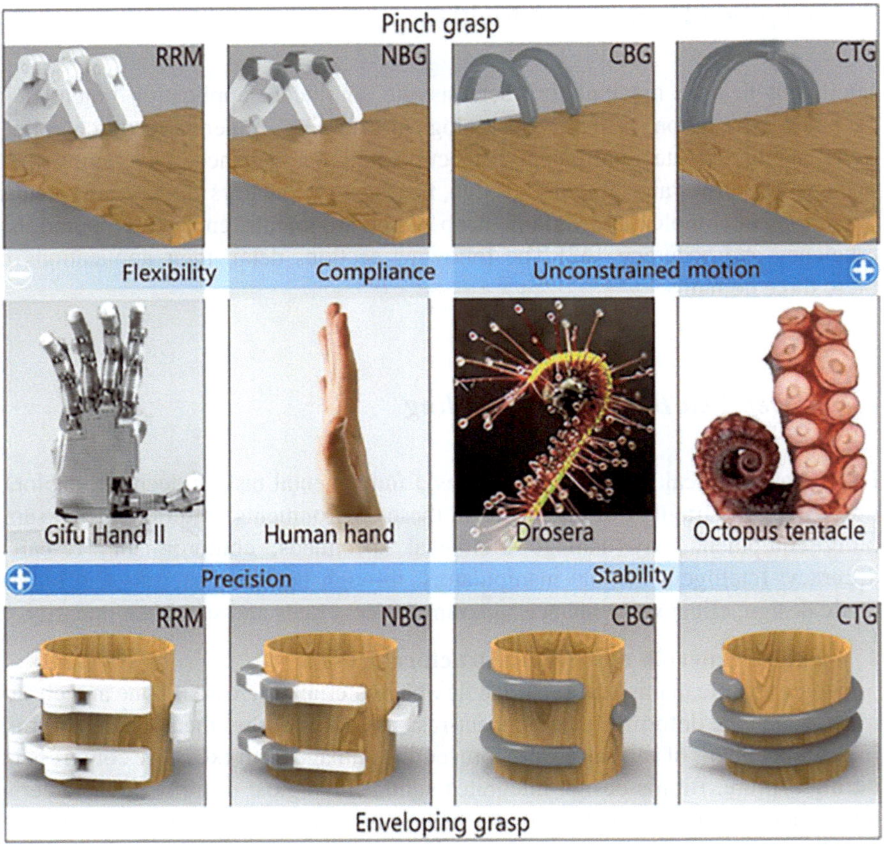

Fig. 8 Comparison of rigid structures and flexible structures

enables better adaptation to complex and variable sampling scenarios, such as narrow deep-sea cavities or areas with strong currents. [45] Simultaneously, the material properties of flexible manipulators ensure more uniform grasping force when contacting biological samples, reducing failure rates and enhancing sample reproducibility.

4.2 Deep-Sea Mineral Resource Collection

Deep-sea mineral resources (such as polymetallic nodules, cobalt crusts, and hydrothermal sulfides) exhibit widespread distribution and high mining value. Traditional rigid equipment often struggles with irregularly shaped and unevenly distributed minerals in deep-sea environments, resulting in low efficiency and potential structural damage. Intelligent flexible manipulators offer a more efficient and adaptable solution for deep-sea mineral resource collection.

Fig. 9 Practical application of a multifunctional soft gripper for deep-sea organism sampling

1. **Adaptive Grasping of Complexly Shaped Minerals**
   The diversity of deep-sea mineral resources dictates their irregular morphologies. Flexible manipulators leverage material deformability to better conform to mineral surfaces, reducing fragmentation or loss due to improper grasping. Bionic suction cup design exemplifies this approach: multiple flexible suction cups on the grasping surface mimic octopus adhesion, enhancing mineral adhesion and stability. The flexible suction cup material automatically adjusts adhesion pressure based on variations in mineral surface roughness and hardness, improving grasping success rates. [46]

2. **Collaborative Operation with Rigid Equipment**
   In deep-sea mining robotic systems, flexible manipulators typically collaborate with rigid grasping equipment. Rigid equipment handles mineral crushing and cutting, while flexible manipulators perform initial grasping and positional adjustments. This collaborative approach effectively avoids operational efficiency losses caused by frequent tool changes and enables flexible switching of grasping strategies based on different mineral characteristics during mining. [47] Additionally, flexible manipulators can perform positional corrections on minerals during grasping, improving the convenience of subsequent transportation and processing.

3. **Multi-mode Integration in Deep-Sea Mining**
   The application of flexible manipulators extends beyond single-task gripping, enabling integration with additional functional modules (such as vibration sensors and water quality monitoring devices) to achieve multifunctional integrated operations. [48, 49] For instance, during deep-sea mineral collection, manipulators can simultaneously perform grabbing and environmental monitoring tasks—such as detecting surrounding water pH levels or metal ion concentrations released by minerals. This multi-mode integration enhances deep-sea mining safety and information acquisition efficiency.

4.3 Inspection and Maintenance of Underwater Equipment

Deep-sea infrastructure (such as subsea pipelines, communication cables, oil and gas wells, etc.) is continuously exposed to extreme environments, making it susceptible to corrosion and physical damage. Therefore, inspection and maintenance of these facilities constitute another critical task in deep-sea exploration. The application of intelligent flexible manipulators in underwater equipment inspection and maintenance is primarily demonstrated through their adaptability to complex environments and precision manipulation capabilities.

1. **Adaptive Operation in Complex Environments**
   Leveraging the deformable properties of their flexible materials, intelligent flexible manipulators can maneuver with agility in confined and curved spaces. Subsea pipelines and cables are typically laid across complex seabed topographies, where traditional rigid inspection equipment often struggles to complete tasks due to size constraints and limited operational angles. Flexible manipulators can perform comprehensive inspections along pipeline or cable surfaces through multi-degree-of-freedom movements like bending and twisting. Integrated sensors—such as ultrasonic and fiber-optic sensors—detect corrosion, cracks, or leaks, enable precise localization and real-time feedback. [50]

2. **Force Feedback Control for Precision Maintenance**
   During underwater equipment maintenance, intelligent flexible manipulators leverage precise force feedback control systems to execute high-precision repair tasks such as replacing seals, removing fouling, or applying anti-corrosion coatings. Flexible materials adapt gripping force to the shape of maintenance targets, preventing equipment damage from excessive force. The force feedback system also optimizes manipulator trajectories based on real-time torque data, ensuring operational stability and safety.

3. **Multimodal Perception in Autonomous Operation**
   The intelligent flexible manipulator integrates multiple perception modalities—including vision, tactile sensing, and force feedback—enabling higher autonomy in complex deep-sea environments. Multimodal perception technology enables the

manipulator to autonomously identify the shape and condition of maintenance targets, dynamically adjusting operational strategies. For instance, upon detecting cracks on a pipeline surface, the manipulator can automatically select appropriate repair tools and autonomously complete crack sealing operations. This autonomous capability significantly enhances deep-sea equipment maintenance efficiency while reducing the frequency of human intervention.

References

1. Lan, X., Shang, L., Zixin, X., et al. (2023). The application and development trends of robots in manufacturing industry [J]. *Technology Information, 21*(24), 44–48.
2. Liu, F., Yang, G., & Yang, K. (2019). Research on flexible manipulators of agricultural picking robots [J]. *Journal of Chinese Agricultural Mechanization and Chemistry, 40*(3), 173–178.
3. Ma, H., Ding, Y., Cui, H., et al. (2025). The application of an intelligent *Agaricus Bisporus*-harvesting device based on FES-YOLOv5s [J]. *Sensors, 25*(2), 519.
4. Chen, M., Chen, F., Zhang, H., Ding, W., & Zhan, C. (2024). Design of flexible picking dexterous hand for ellipsoidal fruits and vegetables [J]. *Journal of Chinese Agricultural Mechanization, 45*(8), 32–37.
5. Abondance, S., Teeple, C. B., Wood, R. J., et al. (2020). A dexterous soft robotic hand for delicate in-hand manipulation [J]. *IEEE Robotics and Automation Letters, 5*(4), 5502–5509.
6. Gao, H., Zhao, Y., Yu, Z., et al. (2024). Fuzzy neural networks-based intelligent tracking control for flexible-joint industry robotic manipulator [C]. In *2024 6th international conference on Industrial Artificial Intelligence (IAI), Shenyang, China* (pp. 1–6).
7. ShiJie Wang, JianFeng Tao, QinCheng Jiang, et al. [J]. Robotics and Computer-Integrated Manufacturing, 2024, 86, 102684.
8. Kefer, M., & Tian, J. (2016). An intelligent robot for flexible quality inspection [C]. In *2016 IEEE International Conference on Information and Automation (ICIA), Ningbo, China* (pp. 80–84).
9. Khan, A., Mineo, C., Dobie, G., et al. (2021). Vision guided robotic inspection for parts in manufacturing and remanufacturing industry [J]. *Journal of Remanufacturing, 11*, 49–70.
10. Nascimento, R., Rocha, C. D., Gonzalez, D. G., et al. (2025). Automated optical system for quality inspection on reflective parts [J]. *The International Journal of Advanced Manufacturing Technology, 137*, 2665–2680.
11. Wang, Y., Xie, Z., Huang, H., et al. (2024). Pioneering healthcare with soft robotic devices: A review [J]. *Smart Med, 3*(1), e20230045.
12. Kong, T., Zheng, Q., Sun, J., et al. (2025). Advances in magnetically controlled medical robotics: A review of actuation systems, continuum designs, and clinical prospects for minimally invasive therapies [J]. *Micromachines (Basel), 16*(5).
13. Fu, S., Chen, B., Li, D., et al. (2023). A magnetically controlled guidewire robot system with steering and propulsion capabilities for vascular interventional surgery [J]. *Advanced Intelligent Systems, 5*(11), 2300267.
14. Zhang, J., Liu, L., Xiang, P., et al. (2024). AI co-pilot bronchoscope robot [J]. *Nature Communications, 15*(1), 241.
15. Dreyfus, R., Boehler, Q., Lyttle, S., et al. Dexterous helical magnetic robot for improved endovascular access [J]. *Science Robotics, 9*(87), eadh0298.
16. Amadeo T, Van Lewen D, Janke T, et al. Soft robotic deployable origami actuators for neurosurgical brain retraction [J]. Frontiers in Robotics and AI, 2022, Volume 8–2021.
17. Zhang, G., Du, F., Xue, S., et al. (2022). Design and modeling of a bio-inspired compound continuum robot for minimally invasive surgery [J]. *Machines, 10*(6).

18. Brancadoro, M., Manti, M., Grani, F., et al. (2019). Toward a variable stiffness surgical manipulator based on fiber jamming transition [J]. *Front Robot AI, 6*, 12.
19. Mitros, Z., Thamo, B., Bergeles, C., et al. (2021). Design and modelling of a continuum robot for distal lung sampling in mechanically ventilated patients in critical care [J]. *Front Robot AI, 8*, 611866.
20. Duan, X., Xie, D., Zhang, R., et al. (2023). A novel robotic bronchoscope system for navigation and biopsy of pulmonary lesions [J]. *Cyborg and Bionic Systems, 4*, 0013.
21. Polygerinos, P., Galloway, K. C., Savage, E., et al. (2015). Soft robotic glove for hand rehabilitation and task specific training. In *Proceedings of the 2015 IEEE International Conference on Robotics and Automation (ICRA), F 26–30 May 2015.*
22. Lee, S.-M., & Park, J. (2024). A soft wearable exoglove for rehabilitation assistance: a novel application of knitted shape-memory alloy as a flexible actuator [J]. *Fashion and Textiles, 11*(1), 14.
23. Wang, X., Cheng, Q., Wang, Z., et al. (2025). *A Pneumatic Soft Glove System Based on Bidirectional Bending Functionality for Rehabilitation [J/OL], 10(3):10.3390/biomimetics10030129.*
24. Yap, H. K., Lim, J. H., Nasrallah, F., et al. (2017). Design and preliminary feasibility study of a soft robotic glove for hand function assistance in stroke survivors [J]. *Frontiers in Neuroscience, 11*, 547.
25. Morris, L., Diteesawat, R. S., Rahman, N., et al. (2023). The-state-of-the-art of soft robotics to assist mobility: A review of physiotherapist and patient identified limitations of current lower-limb exoskeletons and the potential soft-robotic solutions [J]. *Journal of NeuroEngineering and Rehabilitation, 20*(1).
26. Kóra, S., Bíró, A., Prontvai, N., et al. (2024). Investigation of the effectiveness of the Robotic ReStore soft exoskeleton in the development of early mobilization, walking, and coordination of stroke patients: A randomized clinical trial [J]. *Robotics, 13*(3).
27. Kim, J., Lee, G., Heimgartner, R., et al. (2019). Reducing the metabolic rate of walking and running with a versatile, portable exosuit [J]. *Science, 365*(6454), 668–672.
28. Khose, A. S. (2022). Pneumatic exoskeletons for orthopedic rehabilitation of the upper arm—An overview [J]. *Journal of Orthopaedics, Trauma and Rehabilitation, 29*(1), 22104917211056947.
29. Lin, W., Dong, H., Gao, Y., et al. (2025). A systematic review of locomotion assistance exoskeletons: Prototype development and technical challenges [J/OL], *13*(2). https://doi.org/10.3390/technologies13020069
30. Zhang, Y., Li, P., Quan, J., et al. (2023). Progress, challenges, and prospects of soft robotics for space applications [J]. *Advanced Intelligent Systems, 5*(3), 2200071.
31. Hammond, M., Dempsey, A., Ward, W., et al. (2023). A hybrid soft material robotic end-effector for reversible in-space assembly of strut components [J]. *Frontiers in Robotics and AI, 10*, 1099297.
32. Ruiz, F., Arrue, B. C., & Ollero, A. (2024). Thermally-resilient soft gripper for on-orbit operations [C]. In *2024 IEEE/RSJ International Conference on Intelligent Robots and Systems (IROS)* (pp. 14050–14055). IEEE.
33. Jiang, H., Hawkes, E. W., Fuller, C., et al. (2017). A robotic device using gecko-inspired adhesives can grasp and manipulate large objects in microgravity [J]. *Science Robotics, 2*(7), eaan4545.
34. Zhang, Y., Quan, J., Li, P., et al. (2023). A Flytrap-inspired bistable origami-based gripper for rapid active debris removal [J]. *Advanced Intelligent Systems, 5*(7), 2200468.
35. Guest, S. D., & Pellegrino, S. (1992). In extensional wrapping of flat membranes [C]. In *Proceedings of the first international seminar on structural morphology* (p. 25). University of Montpellier.
36. Sabelhaus, A. P., Bruce, J., Caluwaerts, K., et al. (2015). System design and locomotion of SUPERball, an untethered tensegrity robot [C]. In *2015 IEEE International Conference on Robotics and Automation (ICRA)* (pp. 2867–2873). IEEE.

37. Chen, L. H., Kim, K., Tang, E., et al. (2017). Soft spherical tensegrity robot design using rod-centered actuation and control [J]. *Journal of Mechanisms and Robotics, 9*(2), 025001.
38. Peng, R., Wang, Y., Lu, M., et al. (2025). A dexterous and compliant aerial continuum manipulator for cluttered and constrained environments [J]. *Nature Communications, 16*(1), 889.
39. Jahanshahi, H. (2025). A comprehensive review of tactile sensing technologies in space robotics [J]. *Chinese Journal of Aeronautics, 103423.*
40. Wang, Y., Dong, D., Zhou, W., et al. (2025). Flexible wearable device applications for monitoring astronaut health: Current status and challenges [J]. *Wearable Electronics.*
41. Soriano-Amat, M., Fragas-Sánchez, D., Martins, H. F., et al. (2021). Monitoring of a highly flexible aircraft model wing using time-expanded phase-sensitive OTDR [J]. *Sensors, 21*(11), 3766.
42. Wang, Y., Wang, Y., Mushtaq, R. T., & Wei, Q. (2024). Advancements in soft robotics: A comprehensive review on actuation methods. *Materials, and Applications., 16*(8), 1087.
43. Zhou, L., Ren, L., Chen, Y., Niu, S., Han, Z., & Ren, L. (2021). *Bio-Inspired Soft Grippers Based on Impactive Gripping., 8*(9), 2002017.
44. Phillips, B. T., Becker, K. P., Kurumaya, S., Galloway, K. C., Whittredge, G., Vogt, D. M., Teeple, C. B., Rosen, M. H., Pieribone, V. A., Gruber, D. F., & Wood, R. J. (2018). A dexterous, glove-based teleoperable low-power soft robotic arm for delicate deep-sea biological exploration. *Scientific Reports, 8*(1), 14779.
45. Petersen, S., Krätschell, A., Augustin, N., Jamieson, J., Hein, J. R., & Hannington, M. D. (2016). News from the seabed—Geological characteristics and resource potential of deep-sea mineral resources. *Marine Policy, 70*, 175–187.
46. Cheng, W., Zhang, W., Tao, J., Zheng, F., Chu, B., Wang, R., Fang, C., Huai, L., Tao, P., Song, C., Shang, W., Fu, B., & Deng, T. (2024). Octopus-like microstructure of graphene oxide generated through laser–microdroplet interaction for adhesive coating. *ACS Nano, 18*(11), 7877–7889.
47. Li, L., Xie, F., Wang, T., Wang, G., Tian, Y., Jin, T., & Zhang, Q. (2022). Stiffness-tunable soft gripper with soft-rigid hybrid actuation for versatile manipulations. *Soft Robotics, 9*(6), 1108–1119.
48. Díaz, I. M., Pereira, E., Feliu, V., & Cela, J. J. L. (2010). Concurrent design of multimode input shapers and link dynamics for flexible manipulators. *IEEE/ASME Transactions on Mechatronics, 15*(4), 646–651.
49. Wu, M., Afridi, W. H., Wu, J., Afridi, R. H., Wang, K., Zheng, X., Wang, C., & Xie, G. (2024). *Octopus-Inspired Underwater Soft Robotic Gripper with Crawling and Swimming Capabilities., 7*, 0456.
50. Li, G., Chen, X., Zhou, F., Liang, Y., Xiao, Y., Cao, X., Zhang, Z., Zhang, M., Wu, B., Yin, S., Xu, Y., Fan, H., Chen, Z., Song, W., Yang, W., Pan, B., Hou, J., Zou, W., He, S., … Yang, W. (2021). Self-powered soft robot in the Mariana Trench. *Nature, 591*(7848), 66–71.

Technical Challenges and Future Trends

1 Technical Challenges

The rapid evolution of flexible manipulation and intelligent sensing technologies has ushered in transformative capabilities across various domains, including minimally invasive surgery, adaptive manufacturing, human-robot collaboration, and exploration in unstructured environments. From continuum robots navigating through narrow spaces to soft grippers handling delicate objects with human-level dexterity, these systems epitomize the convergence of advanced materials, distributed sensing, and adaptive control. Yet, despite significant progress chronicled in preceding chapters of this book, the path toward robust, scalable, and autonomous deployment remains fraught with profound technical barriers.

This final chapter shifts focus from established methodologies and successes to the unsolved challenges that define the frontier of this field. While earlier sections detailed the principles, designs, and applications of flexible manipulators and sensing systems, we now confront the critical limitations that hinder their reliability, performance, and broader adoption. Addressing these challenges is not merely an academic exercise; it is an imperative for unlocking the next generation of intelligent machines capable of operating safely and effectively in dynamic, real-world settings—where uncertainty prevails and perfection is unattainable.

The significance of this discussion extends beyond immediate technical hurdles. Breakthroughs in modeling, control, sensing, actuation, and integration will catalyze innovations in areas as diverse as personalized healthcare, sustainable manufacturing, and disaster response. Conversely, unresolved challenges risk stagnating progress, limiting the application of these technologies to controlled laboratory environments or niche applications.

This chapter systematically dissects these bottlenecks across four interconnected domains: modeling and simulation, control and planning, sensing and perception, and finally, material and actuation. Our goal is twofold: to provide researchers and

J. Qu, *Robotic Flexible Manipulation and Intelligent Sensing Technologies*,
https://doi.org/10.1007/978-981-95-6232-9_6

engineers with a clear road map of critical gaps that demand attention and to inspire collaborative, interdisciplinary efforts to overcome them.

1.1 Modeling and Simulation

Accurate modeling and simulation of flexible manipulation systems form the foundational bedrock for design, control, and deployment. Yet, the intrinsic compliance and large-deformation capabilities of these systems—ranging from continuum surgical robots to soft grippers—introduce complexities that defy conventional rigid-body paradigms. At the core lies the **nonlinear dynamics of deformable structures**, where geometric deflections, material hyperelasticity or viscoelasticity, and dynamic inertial coupling conspire to render classical Lagrangian formulations inadequate. While Cosserat rod model [1] and Absolute Nodal Coordinate Formulation (ANCF) offer advanced frameworks [2], their computational intensity often precludes real-time application, especially in systems with high degrees of freedom [3].

The challenge escalates dramatically when object **interactions** are taken into account. Predicting contact mechanics between a soft manipulator and unstructured surfaces—be it biological tissue or irregular industrial objects—remains fraught with uncertainty. Unknown contact points, sliding dynamics, and force-deformation coupling create ill-posed inverse problems [4]. Data-driven approaches have emerged to model heterogeneous substrates, such as granular media or living tissue [5], yet they require extensive validation and struggle to generalize across various scenarios. This ambiguity directly compromises tasks requiring precision, such as delicate object manipulation or minimally invasive surgery.

Further complicating the landscape is the **coupling of multiple physics**. Flexible systems frequently operate under intertwined physical domains: fluid-structure interactions (FSI) in underwater soft robots or pneumatic actuators [6], thermo-mechanical hysteresis in shape memory alloys [7], or electromechanical effects in dielectric elastomers [8]. Traditional sequential solvers decouple these phenomena at the cost of accuracy, whereas monolithic approaches impose unacceptable computational loads, making them impractical for control or design iteration.

This exposes the critical tension between **simulation fidelity and real-time feasibility**. High-fidelity methods, such as finite element analysis (FEA) or detailed Cosserat simulations, offer precision but operate orders of magnitude slower than physical systems [9]. Reduced-order models, including proper orthogonal decomposition or physics-informed neural networks (PINNs) [10], accelerate computation but sacrifice robustness. They often fail to adapt to novel configurations, dynamic events, or parametric variations, which limits their utility in adaptive control loops.

Ultimately, **irreducible uncertainties** pervade these models. Manufacturing tolerances, material batch variations, and environmental drift introduce parametric noise, while unmodeled effects, such as tendon friction or pneumatic leakage, create latent disturbances [11]. Although Bayesian filters or Gaussian processes can compensate [12], they rely on dense sensor networks and introduce computational overhead, highlighting a fundamental trade-off between model complexity and operational practicality.

Emerging hybrid approaches—**blending first-principles physics with data-driven corrections**—are one possible direction. For instance, real-time Cosserat solvers augmented with online parameter identification, or PINNs enforcing thermodynamic constraints on SMA actuators, point toward adaptive, computationally tractable frameworks. However, standardizing benchmarks for deformable system simulation remains critical to accelerate progress. As exemplified in robotic laparoscopic surgery, where simulating organ-tool interactions demands unrealizable FSI-contact coupling [13], overcoming these barriers will unlock safer, more autonomous flexible systems.

1.2 Control and Planning

The transition from rigid to flexible manipulators fundamentally disrupts traditional control paradigms. Where conventional robots leverage precise kinematics and deterministic dynamics, compliant structures—characterized by distributed deformation, underactuation, and nonlinear hysteresis—demand controllers that reconcile adaptability with stability in the face of intrinsic uncertainty. **Model-based approaches**, while theoretically grounded, falter when confronted with the reality of imperfect representations. As detailed before, inaccuracies in simulating hyperelasticity, contact forces, or multi-physics coupling propagate directly into control loops. Model predictive control (MPC), for instance, relies critically on real-time deformation forecasts to optimize trajectories [4]. Yet, computational delays or unmodeled disturbances—such as tendon friction in continuum robots or air leakage in pneumatic actuators—can degrade performance, causing oscillations or task failure in delicate operations like tissue retraction or micro-assembly [11].

This fragility amplifies when pursuing real-time shape and force regulation. Achieving millimeter-scale tip positioning in a cable-driven endoscope or maintaining sub-Newton contact forces during the manipulation of fragile objects requires controllers that compensate for dynamic coupling between segments and environmental compliance. While impedance and admittance control frameworks offer partial solutions, they struggle with the dual challenge of large, rapid deformations and fine force modulation—especially when sensor feedback is sparse or noisy [14]. The problem is exemplified in soft grippers handling irregular produce: geometric uncertainties in object shape, combined with material viscoelastic relaxation, necessitate continuous gain scheduling beyond classical PID capabilities.

Consequently, learning-based methods—reinforcement learning (RL) and imitation learning (IL)—have surged as alternatives, promising adaptability through data-driven policy optimization. However, these approaches confront steep barriers in flexible systems. **Sample inefficiency** plagues RL, as physical experiments on deformable structures are slow and risk damage [15]. Sim-to-real transfer remains problematic due to the "reality gap" in soft dynamics simulation [16]. More critically, **safety and robustness** are often compromised: black-box neural network policies lack interpretability, may ignore force constraints during exploration, and fail catastrophically under novel conditions—a prohibitive risk in human-in-the-loop scenarios like rehabilitation robotics.

These limitations extend into the realm of **motion planning and autonomy**. Planning collision-free paths for hyper-redundant continuum manipulators in confined spaces involves navigating high-dimensional configuration spaces with complex constraints [17]. While sampling-based planners handle obstacles, they disregard deformation energy minimization or dynamic stability, leading to inefficient or unstable motions. Meanwhile, enabling true autonomy—where systems self-adapt to tool damage, payload shifts, or environmental changes—demands online re-planning capabilities currently hindered by computational latency and incomplete world models [18]. Even teleoperation, a stopgap for complex tasks, grapples with haptic transparency: transmitting realistic force feedback through flexible interfaces requires low-latency, distortion-free control loops rarely achievable with current soft sensor-controller co-design [19].

Promising pathways include **hybrid neuro-symbolic architectures** blending RL's adaptability with physics-based safety filters, **online model identification** to refine controllers under drift, and **topology-aware planners** optimizing for deformation energy as showcased in underwater soft robotics—where currents, buoyancy shifts, and thruster hysteresis challenge conventional control—overcoming these hurdles demands co-evolution of modeling, sensing, and control theory.

1.3 Sensing and Perception

Intelligent sensing is the linchpin enabling flexible systems to interpret deformation, contact, and environmental states—yet the very compliance that defines these systems creates **fundamental paradoxes for perception**. Traditional sensing paradigms, designed for rigid structures, falter when subjected to continuous curvature changes, large strains, or dynamic material properties. **Distributed sensing networks** embedded in soft actuators or continuum manipulators face **integration-compromise trilemmas**: achieving high spatial resolution, mechanical robustness, and signal stability simultaneously proves elusive. For instance, optical fiber sensors (FBGs) offer millimeter-scale strain mapping but fracture at large elongations [20]; liquid-metal circuits have significant deformable ability yet drift with motion artifacts [21]. This forces designers into trade-offs that limit situational awareness, as seen in surgical robots, where incomplete shape feedback risks tissue perforation.

The challenge intensifies in **multimodal sensor fusion**. Flexible systems demand a cohesive interpretation of heterogeneous data streams—tactile arrays measuring pressure distribution, inertial units tracking segment orientation, cameras capturing global deformation, and microphones detecting slip events. However, **spatiotemporal misalignment** arises from differing sensor bandwidths and placements: strain gauges update at kHz rates, while vision processes usually occur at 30 Hz, creating temporal gaps during fast bending [22]. Spatial calibration errors compound when sensors shift relative to deformable substrates. Consequently, Bayesian filters or deep fusion networks struggle to reconstruct accurate state estimates from conflicting inputs, particularly in occlusion-prone environments like confined industrial machinery or cluttered surgical fields [23].

Environmental perception further exposes fragility. Flexible manipulators operating in unstructured settings—harvesting fruits in orchards, inspecting pipe interiors, or navigating disaster rubble—confront **signal degradation** from real-world perturbations. Visual occlusion by self-contact or environmental clutter, specular reflections on wet surfaces, and electromagnetic interference in metal-rich facilities corrupt raw data. While learning-based approaches mitigate some issues [24], they lack robustness to **domain shifts**: a vision model trained in controlled labs fails when dust diffuses surgical field lighting or when grape texture varies seasonally. This necessitates costly retraining or hardware redundancy, undermining deployment scalability.

At the hardware-software interface, **computational constraints** create critical bottlenecks. High-density sensor arrays generate data streams exceeding 1 Gbps, demanding edge processing for real-time feedback. Current embedded platforms cannot concurrently run physics-based state estimation and deep perception pipelines under relatively low power budgets [25]. This forces approximations: downsampling tactile data loses micro-slip signatures critical for grasp stability, while quantized neural networks reduce shape reconstruction accuracy by >20% [26]—compromising control precision precisely when flexibility promises adaptive advantage.

Finally, **long-term reliability** remains largely unaddressed. Cyclic deformation induces **material fatigue** in stretchable conductors, while environmental exposure, including humidity, chemicals, and sterilization, further accelerates **sensor drift**. Recalibration in situ is impractical during tasks like autonomous underwater inspection. Although self-diagnostic algorithms detect anomalies [27], they cannot compensate for irreversible degradation—gradually eroding system trustworthiness.

1.4 Material and Actuation

The promise of flexible manipulation hinges on advanced materials and actuators that transcend conventional rigidity—yet these enabling technologies confront **intrinsic performance ceilings** that propagate through every layer of system design. **Multifunctional material requirements** present a fundamental trade-off: high elasticity for compliant deformation often sacrifices tensile strength, while electrical conductivity typically degrades under cyclic loading [28]. Even state-of-the-art composites—such as nanoparticle-reinforced silicones or liquid metal-embedded elastomers—struggle to simultaneously deliver biocompatibility, self-healing capability, and environmental resilience [29].

Durability under operational stress emerges as a critical frontier. Repeated large-strain deformation induces **microstructural fatigue** in elastomers: chain scission in polymers, delamination in stretchable electrodes, or void nucleation in pneumatic chambers [30]. Environmental aggressors accelerate degradation—UV radiation embrittles silicones in outdoor applications, body fluids swell hydrogels in medical robots, and industrial chemicals plasticize polyurethanes. While self-healing materials partially mitigate damage [31], their healing cycles are finite, and activation requires heat/light inaccessible in confined spaces. Consequently, field-deployable systems like disaster rescue robots face unpredictable lifespan cliffs, where a single cut or abrasion disables critical segments.

The **control-compounding effects** of these limitations are profound. For example, hysteresis in SMA actuators introduces position errors without sensor feedback [32], forcing reliance on fragile strain gauges. Creep in viscoelastic tendons causes grasp-force relaxation in soft grippers during prolonged holds [33], risking object drops. Moreover, **thermal management** in densely integrated systems—where Joule heating from embedded electronics combines with actuator losses—can trigger thermal runaway, degrading adjacent sensors or structural integrity [34]. This interdependence demands co-design approaches currently hindered by disciplinary silos: materials scientists optimize strain tolerance, while control engineers battle hysteresis-induced instability.

2 Ethical and Social Impacts

Flexible manipulators, with deformable soft materials, compliant mechanisms, and force-position hybrid control at their core, can well replicate the appearance and capabilities of humans or animals and can complete safe, compliant, and highly adaptable tasks in narrow, uncertain, and close-contact scenarios with the human body. Its advantages over traditional rigid manipulators lie in gentle contact, shape reconstruction, and lightweight, but it also comes with engineering challenges, such as strong nonlinear modeling problems, complex sensor-control coupling, material fatigue, and maintenance costs. With its accelerated application in fields, such as medical-assisted surgery, elderly care, agricultural picking, and service robots, its moral and social impacts are gradually becoming prominent: safety and trust have become the prerequisites for deployment. Automation reshapes the labor market and involves social equity in employment opportunities and distribution. In the field of assistive and rehabilitation, it is necessary to take into account therapeutic effects, informed consent, and the protection of vulnerable groups. Data-driven intelligence brings about ethical risks of privacy, bias, and explainability; Material selection, energy consumption, and full life cycle management are related to environmental sustainability issues. Based on this, this chapter elaborates on the moral and social impacts of flexible manipulators from the following five aspects:

1. **Safety and Trust**

As flexible mechanical hands gradually enter scenarios close to the human body such as manufacturing, medical care, and nursing, the topic of "safety and trust" is a necessary condition for being accepted by society. Compared with rigid systems, flexible materials and compliant structures can significantly reduce the damage caused by impact energy and contact, but they also introduce uncertainties such as strong nonlinearity, hysteresis, and material time-varying, leading to the predicament of "seeming safer but more difficult to prove safe." Therefore, safety should not be judged directly by the "softness or hardness" of materials but must be supported by verifiable engineering evidence and institutionalized governance. Trust should not be established based on favorable impressions or personified appearances, but rather on performance stability, transparency, and accountability mechanisms.

The main hazards of flexible arms include: sudden failure of the structure during operation due to contact squeezing, overload tearing, and material fatigue. Sensor data drift, model mismatch, and limit failure in force-position hybrid control; Communication delay, network attacks, and software regression risks in upgrade rollback; And human errors caused by excessive trust and misinterpretation of the situation. For example, the FDA MAUDE of the United States has reported multiple cases that during the operation of the da Vinci Xi surgical system flexible EndoWrist Stapler 30, the silicone sheath at the end of the operation was torn and left in the human body, causing secondary losses (delayed suturing and additional anesthesia); The Dutch Labour Inspection Agency's "Ongevallenrapport 2019–112" reported that a company's three-finger pneumatic soft gripper experienced a fatigue rupture in the inflation cavity during maintenance, and the fragments hit the operator's cornea, causing ocular surface trauma. BG ETEM Accident Report: At the Dingolfing plant of BMW Germany, when using the flexible silicone end system independently developed by KUKA LBR iiwa +, a control race occurred when switching from manual teaching mode to automatic. The flexible joint at the wrist "locked," instantly injuring the back of the worker's hand and causing a bone fracture. China Medical Device Information reported that in a rehabilitation hospital in Shenzhen, China, when using soft pneumatic hand rehabilitation gloves, the upper limit parameter of pressure control was reset after the firmware upgrade, and the inflation was greater than 90 kPa. The patient's right fingertip was ischemic for 12 min, resulting in grade one soft tissue injury.

When using flexible arms, the design principle of "safety first" should be adhered to in engineering. For instance, the risk at the source can be reduced by adopting fatigue-resistant and tear-resistant materials and optimizing the energy dissipation of the compliant mechanism. For end-effectors and replaceable fixtures, predictable failure modes and mechanical limits should be introduced to prevent software anomalies from amplifying the risk. For medical and rehabilitation applications, it is also necessary to consider the material performance degradation and biocompatibility boundaries caused by sterilization cycles, establish a performance reverification process after accelerated aging and cycle sterilization, and strictly ensure safety. In the perception and control layer of flexible manipulators, multi-mode redundancy and cross-checking are emphasized. For instance, the integration of force, vision, and touch enhances the detectability of contact events. Constraint-based safety control (such as speed and separation monitoring, contact limiting) and online confidence assessment achieve adaptive conservative strategies, and key safety attributes can be verified and simulated in advance. Fault safety and passive protection provide the last line of defense: an independent emergency stop circuit, energy isolation, power-off self-stabilization, and mechanical safety design, ensuring that the system tends to a low-risk state when the sensing or control layer fails. For manual close collaboration, soft wrapping, rounded corner design, and compressible end-effectors should be configured, and the maximum contact force-time curve related to the pain threshold should be measured and verified for compliance.

In product implementation, to avoid insufficient trust, an explainable human-machine interface should be provided, showing action intentions, force thresholds, environmental understanding confidence levels and residual risks, ensuring behavioral consistency and predictable trajectories, clearly defining permissions, takeover conditions and responsibility boundaries, and presenting the logic of abnormal handling in an understandable way. For clinical and elderly care users, informed consent forms should be implemented. For frontline operators, situational drills and misuse prevention education should be carried out to avoid automation errors and alarm fatigue.

2. **Labor Market and Social Equity**

Flexible manipulators, with their soft materials and friendly human-machine collaboration, are entering high-contact and high-service scenarios, such as assembly, sorting, medical rehabilitation, and home care. Compared with traditional industrial robots, its technical path is more inclined toward "replacing micro-skills, enhancing physical strength, and reconfiguring processes," thus providing unique directional guidance for the impact on employment structure, income distribution, and opportunity acquisition.

Flexible manipulators can stably perform repetitive and fine actions in unstructured environments. They are the first to replace intermediate-skilled positions that rely on experience but can be process-oriented (such as fine packaging, trim assembly, and simple grasping). For instance, when the Da Vinci surgical robot is combined with a flexible mechanical hand, it can replace human assistants, reduce the operational errors caused by traditional manual surgeries in surgery, and achieve more precise surgeries or direct remote operations in cavities or complex, deep, and narrow areas, significantly improving the quality of surgical operations. At the same time, it places higher demands on skills in system integration, perception calibration, maintenance and operation, and human-machine collaborative design, giving rise to high-skilled positions such as materials science, robot programming, and embedded system design. This brings new opportunities to well-trained technical practitioners and creates a polarized pressure of "middle squeezing." Without targeted skills transformation response policies, young newcomers to the industry and middle-aged and elderly workers with low educational attainment will face problems such as structural unemployment and wage stagnation. Re-employment and humanitarian care should be provided to groups affected by job transfer through mechanisms such as education, skills training, and social security to prevent further increase in social inequality.

Flexible mechanical hands are a technology with high fixed asset and strong intellectual property intensive nature. The equipment, consumables, and software subscriptions form a "lock-in effect," which may drive the industry and service supply to concentrate on a few manufacturers and leading medical institutions. The cost structure of hospitals has shifted from "human resources + general equipment" to "capital expenditure + specialized consumables." If the payment end does not fully recognize the effectiveness in the short term, the financial pressure will be passed on to the performance of departments and individuals, changing the income composition and assessment logic of medical staff. The trend of platformization is conducive to economies of scale and quality consistency, but it may also create a "case siphon," intensifying the Matthew effect among regions and hospital grades. For clinical practitioners, the technical premium may rise in the short term, but as supply expands and paths become standardized, the individual premium tends to return. Team collaboration, data governance, and quality results will become new salary anchors. At the policy level, interoperability standards, consumable price comparisons, and transparent cost accounting should be adopted to weaken lock-in and prevent excessive concentration of competition policies. And promote "value-based payment," converting real benefits such as reduced complications and shortened hospital stays into reasonable payments, to avoid disproportionately passing on the cost of innovation to patients and frontline staff.

3. **The Impact in the Field of Assistive and Rehabilitation**

In rehabilitation, elderly care, and disability assistance scenarios, flexible manipulators, with their low-stiffness materials and variable compliance control, can perform fine and safe movements without harming the human body. Therefore, they are regarded as a key technology for enhancing self-care ability and quality of life. For instance, Harvard University's soft pneumatic gloves, driven by airbags and silk threads, provide "on-demand assistance" grasping operations for patients with cerebral stroke. They can help them button up and zip up independently without inhibiting the recovery of residual muscle strength. The MySpoon dining robot from Itochu, Japan, uses soft gripper claws and visual tracking to automatically adjust the spoon angle after recognizing the shape of the food, reducing misfeeding and choking. The Rewalk-Soft lightweight exoskeleton from South Korea, combined with electromyography sensing, provides the shoulder and elbow support needed by elderly people with muscle weakness when dressing. In higher-precision medical operations, although the Da Vinci surgical robot adopts a rigid mechanism, it also relies on compliant control algorithms and force feedback, enabling surgeons to perform minimally invasive operations such as suturing blood vessels or stripping nerves as if they were "wearing a flexible mechanical hand." With the maturation of stretchable sensors, flexible circuits, and safety algorithms, these flexible hand systems are gradually moving out of laboratories and into care institutions and home environments, providing reliable human-machine collaborative solutions for the entire chain of "assistance—rehabilitation—care."

Flexible manipulators often appear in the fields of assistive and rehabilitation in the form of exoskeletons, soft gloves, swallowing/gait assistance, etc., directly altering daily living functions. On the one hand, it restores abilities such as picking up objects, walking, and eating by conforming to the variable stiffness and safety compliance of the human body, thereby enhancing a person's independence. On the other hand, it also brings the risk of "learned non-use"—if over-assisted for a long time, residual muscle strength and neural plasticity may be inhibited, forming a new type of dependence. Ethically, the principles of "minimum adequate assistance" and "gradual withdrawal of assistance" should be adhered to, and assistance should be dynamically adjusted to encourage active participation. In terms of safety, it is necessary to ensure physical protection against skin cuts, pressure injuries, and out-of-range joints and provide understandable and explainable control modes as well as mechanical escape designs that can be removed at any time.

Considering the issues of cost and accessibility, if there is a lack of government subsidies, public welfare support, or inclusive policies, the wide application of flexible mechanical hands in the medical and auxiliary fields may mainly be concentrated among groups with better economic conditions, leading to a further expansion of the technological gap. For instance, in some developed countries in Europe and America, with the support of the government and insurance institutions, the elderly and people with disabilities can obtain government-subsidized intelligent assistive devices, such as flexible mechanical hands or exoskeleton systems, thereby improving their quality of life. In developing countries or low-income regions, due to the high costs of equipment procurement and maintenance, these advanced technologies may only be used by a few wealthy families or institutions, making it difficult for ordinary patients to afford them, thus creating a "digital divide" and "health inequality." This requires the joint efforts of policymakers, public welfare organizations, and the industrial sector. Through measures such as subsidies, price control, and technological innovation, costs should be reduced to ensure the fair popularization of flexible manipulator technology and prevent further aggravation of social inequality.

4. **Data Privacy and Ethical Risks**

Many flexible mechanical hands need to obtain user information (such as biological signals, movement trajectories, and living habits) in real-time based on multiple sensors or visual systems in medical rehabilitation, daily care, and disability assistance services, so as to achieve precise flexible grasping or assist actions. During this process, if the collection, storage, and transmission of personal data lack necessary data encryption, access control, and compliance review, there may be a risk of privacy leakage.

For instance, the medical privacy issues arising from continuous sensing and home wearing are equally prominent: data such as heart rate, gait, and audio and video of the home environment may be used to infer health status and lifestyle. Therefore, it is necessary to minimize collection, edge computing, explicitly state the purpose and the right of withdrawal, and adopt hierarchical consent and joint decision-making by guardians in the elderly and cognitively impaired populations.

In medical, industrial collaboration, and home rehabilitation scenarios, flexible manipulators continuously collect highly sensitive data, such as grip strength curves, electromyographic signals, skin temperature and humidity, movement patterns, positions, and audio and video clips, and are often associated with medical records, insurance, and identity information. The risk lies not only in single-point leakage but also in the data concatenation and re-identification across devices and platforms. Once used for profiling, differential pricing, or employment screening, it will cause long-term and irreversible harm to individuals. The general principle should be "minimum necessity + purpose limitation," with default end-to-end processing and pseudonymized storage. Key links should adopt end-to-end encryption, hardware root trust, and key domain management. Or establish fine-grained authorization, withdrawable consent, data portability and deletion rights, and incorporate access control, log auditing, and abnormal alerts as part of daily operation and maintenance. Before deployment, a privacy impact assessment and threat modeling should be completed, covering scenarios such as sensor out-of-control, cloud abuse, and unauthorized third-party SDKS. For children, the elderly, and users with cognitive limitations, enhanced notification and guardianship consent should be provided, and at the UI level, interactions with "privacy as default" should be adopted to reduce unintentional exposure.

5. **Environmental Sustainability**

While traditional flexible robot technology has brought convenience to humans, it has also caused serious environmental burdens, such as the accumulation of electronic waste, toxic materials, and energy consumption. Therefore, integrating the concept of sustainable development into the design of flexible robots is not only an inevitable trend of technological progress but also an important strategy to address the global environmental crisis.

In terms of materials, traditional soft robots often use materials such as silicone rubber, plastic, and metal. Although they have excellent performance, they have problems such as environmental pollution and difficulty in recycling. According to statistics, the global electronic waste is increasing at a rate of 100,000 tons per day. While bringing convenience to human beings, it also brings a heavy price to environmental pollution. To solve this problem, the key lies in developing degradable and multifunctional materials and practicing the concept of sustainable development, such as elastomers, polysaccharides, and bioplastics. In terms of manufacturing, low-energy consumption and low-waste technologies such as 3D printing and laser cutting can be employed. In terms of energy supply, how to solve the problems of heavy metal pollution and insufficient energy supply in traditional batteries is an important challenge for achieving sustainable flexible robots. In the future, degradable batteries, solar films, and the like can be used.

The sustainable development of flexible robots requires interdisciplinary collaboration, integrating materials science, biology, engineering, and ecology. In the future, flexible robots should coexist harmoniously with the environment like natural creatures, achieving a paradigm shift from "manufacturing" to "growth." By promoting the concept of green design, flexible robots will play a greater role in fields such as healthcare, agriculture, and disaster relief, while minimizing their impact on the Earth's ecology.

3 Perspectives

As cutting-edge fields such as artificial intelligence, big data, and biotechnology continue to emerge, the ethical risks and social changes they bring have gradually come to the fore. While technological innovation can greatly promote social progress and improve people's lives, it also raises the question of how to balance the contradictions between efficiency and fairness, privacy and sharing, and innovation and security.

The widespread application of big data and cloud computing makes personal information extremely easy to be collected and stored, posing unprecedented challenges to citizens' privacy. How to promote data sharing and information dissemination while preventing abuse and leakage requires relevant departments and enterprises to fully consider ethical bottom lines when formulating regulations and technical solutions. At the same time, in the field of artificial intelligence, the transparency, fairness of algorithms, and the timely correction of biases have also become important issues that need to be addressed urgently. Ensuring that technological decisions do not lead to social injustice due to data bias is the top priority in current technological ethics construction.

The popularization of the Internet and smart devices not only improves the convenience of information access but also, to a certain extent, narrows the geographical and temporal distances. However, this transformation also poses the risk of marginalizing certain groups when they adapt to new technologies. For example, the digital divide is particularly evident among different age, regional, and economic groups. If the benefits of technology cannot be shared by the general public, it may exacerbate the uneven distribution of social resources and cause new social conflicts and instability factors.

Therefore, all sectors should jointly promote the construction of a diverse and inclusive technological development environment. Governments, enterprises, and social organizations need to strengthen communication and cooperation, formulate transparent, responsible, and forward-looking policies and measures and promote technological innovation while protecting citizens' basic rights. At the same time, through public participation and extensive discussions, technological ethics and social impact issues should be brought into the public eye, forming a benign mechanism of joint supervision and implementation by the entire society.

References

1. Antman, S. S. (2005). *Nonlinear problems of elasticity*. Springer.
2. Shabana, A. A. (2020). *Dynamics of multibody systems*. Cambridge University Press.
3. Renda, F., Boyer, F., Dias, J., & Seneviratne, L. (2018). Discrete Cosserat approach for multi-section soft manipulator dynamics. *IEEE Transactions on Robotics, 34*(6), 1518–1533.
4. Duriez, C. (2013, May). Control of elastic soft robots based on real-time finite element method. In *2013 IEEE international conference on robotics and automation* (pp. 3982–3987). IEEE.
5. Verheyen, C. A., Uzel, S. G., Kurum, A., Roche, E. T., & Lewis, J. A. (2023). Integrated data-driven modeling and experimental optimization of granular hydrogel matrices. *Matter, 6*(3), 1015–1036.
6. Lin, Z., Hess, A., Yu, Z., Cai, S., & Gao, T. (2019). A fluid–structure interaction study of soft robotic swimmer using a fictitious domain/active-strain method. *Journal of Computational Physics, 376*, 1138–1155.
7. Lagoudas, D. C., & Lagoudas, D. C. (2008). *Shape memory alloys* (Vol. 1). Springer.
8. Suo, Z. (2010). Theory of dielectric elastomers. *Acta Mechanica Solida Sinica, 23*(6), 549–578.
9. Falkenhahn, V., Mahl, T., Hildebrandt, A., Neumann, R., & Sawodny, O. (2015). Dynamic modeling of bellows-actuated continuum robots using the Euler–Lagrange formalism. *IEEE Transactions on Robotics, 31*(6), 1483–1496.
10. Cuomo, S., Di Cola, V. S., Giampaolo, F., Rozza, G., Raissi, M., & Piccialli, F. (2022). Scientific machine learning through physics–informed neural networks: Where we are and what's next. *Journal of Scientific Computing, 92*(3), 88.
11. Kato, T., Okumura, I., Kose, H., Takagi, K., & Hata, N. (2016). Tendon-driven continuum robot for neuroendoscopy: Validation of extended kinematic mapping for hysteresis operation. *International Journal of Computer Assisted Radiology and Surgery, 11*(4), 589–602.
12. Hu, J., Hu, W., Cao, D., Li, S., Chen, J., Huang, Y., et al. (2024). Robust multiarea distribution system state estimation based on structure-informed graphic network and multitask Gaussian process. *IEEE Transactions on Industrial Informatics, 20*(8), 10599–10612.
13. Zhang, X., Wang, Z., Sun, W., & Mukherjee, M. (2024). Real-time non-uniform surface refinement model for lung adenocarcinoma surgery. *Medical & Biological Engineering & Computing, 62*(1), 183–193.
14. Thieffry, M. (2019). Model-based dynamic control of soft robots. Doctoral dissertation, Université Polytechnique des Hauts-de-France.
15. Bhagat, S., Banerjee, H., Ho Tse, Z. T., & Ren, H. (2019). Deep reinforcement learning for soft, flexible robots: Brief review with impending challenges. *Robotics, 8*(1), 4.
16. Gao, J., Michelis, M. Y., Spielberg, A., & Katzschmann, R. K. (2024). Sim-to-real of soft robots with learned residual physics. *IEEE Robotics and Automation Letters., 9*, 8523.
17. Franco, E., & Garriga-Casanovas, A. (2021). Energy-shaping control of soft continuum manipulators with in-plane disturbances. *The International Journal of Robotics Research, 40*(1), 236–255.
18. Webster, R. J., III, & Jones, B. A. (2010). Design and kinematic modeling of constant curvature continuum robots: A review. *The International Journal of Robotics Research, 29*(13), 1661–1683.
19. Selim, M., Dresscher, D., & Abayazid, M. (2024). Virtual needle insertion with enhanced haptic feedback for guidance and needle–tissue interaction forces. *Sensors, 24*(17), 5560.
20. Hao, J., Zhang, Z., Wang, S., & Shi, C. (2023). 2D shape estimation of a pneumatic-driven soft finger with a large bending angle based on learning from two sensing modalities. *Advanced Intelligent Systems, 5*(10), 2200324.
21. Zhuang, Q., Yao, K., Wu, M., Lei, Z., Chen, F., Li, J., et al. (2023). Wafer-patterned, permeable, and stretchable liquid metal microelectrodes for implantable bioelectronics with chronic biocompatibility. *Science Advances, 9*(22), eadg8602.
22. Taunyazov, T., Sng, W., See, H. H., Lim, B., Kuan, J., Ansari, A. F., et al. (2020). Event-driven visual-tactile sensing and learning for robots. *arXiv preprint arXiv:2009.07083*.

23. Chen, B., Chin, T. J., & Klimavicius, M. (2022). Occlusion-robust object pose estimation with holistic representation. In *Proceedings of the IEEE/CVF winter conference on applications of computer vision* (pp. 2929–2939).
24. Yan, M., Zhu, Y., Jin, N., & Bohg, J. (2020). Self-supervised learning of state estimation for manipulating deformable linear objects. *IEEE Robotics and Automation Letters, 5*(2), 2372–2379.
25. Zhang, H., Wang, Q., Feng, J., Yang, Z., & Wang, H. (2021). Design of environmental perception system based on edge computing. *Journal of Computers, 32*(2), 233–239.
26. Hubara, I., Courbariaux, M., Soudry, D., El-Yaniv, R., & Bengio, Y. (2018). Quantized neural networks: Training neural networks with low precision weights and activations. *Journal of Machine Learning Research, 18*(187), 1–30.
27. Zhu, M., Li, J., Wang, W., & Chen, D. (2021). Self-detection and self-diagnosis methods for sensors in intelligent integrated sensing system. *IEEE Sensors Journal, 21*(17), 19247–19254.
28. Ouabdesselam, M., & Hudson, P. J. (1991, October). An investigation of the effect of cyclic loading on fracture conductivity. In *SPE annual technical conference and exhibition?* (p. SPE-22850). SPE.
29. Liu, J., Yao, Y., Chen, S., Li, X., & Zhang, Z. (2021). A new nanoparticle-reinforced silicone rubber composite integrating high strength and strong adhesion. *Composites Part A: Applied Science and Manufacturing, 151*, 106645.
30. Li, C., Yang, H., Suo, Z., & Tang, J. (2020). Fatigue-resistant elastomers. *Journal of the Mechanics and Physics of Solids, 134*, 103751.
31. Gomez, E. F., Wanasinghe, S. V., Flynn, A. E., Dodo, O. J., Sparks, J. L., Baldwin, L. A., et al. (2021). 3D-printed self-healing elastomers for modular soft robotics. *ACS Applied Materials & Interfaces, 13*(24), 28870–28877.
32. Mohan, S., & Banerjee, A. (2021). Modelling of minor hysteresis loop of shape memory alloy wire actuator and its application in self-sensing. *Smart Materials and Structures, 30*(5), 055011.
33. Paul Buckley, C., Samuel Salisbury, S. T., & Zavatsky, A. B. (2016). Viscoelasticity of tendons under transverse compression. *Journal of Biomechanical Engineering, 138*(10), 101004.
34. Ju, Y. S. (2022). Thermal management and control of wearable devices. *Iscience, 25*(7), 104587.

MIX
Papier aus verantwortungsvollen Quellen
Paper from responsible sources
FSC® C105338

If you have any concerns about our products,
you can contact us on
ProductSafety@springernature.com

In case Publisher is established outside the EU,
the EU authorized representative is:
Springer Nature Customer Service Center GmbH
Europaplatz 3, 69115 Heidelberg, Germany

Printed by Libri Plureos GmbH
in Hamburg, Germany